4-17-79

Energy and the Community

ENERGY AND THE COMMUNITY presents the results of a colloquium, "Energy and Patterns of Human Settlement", convened by the Center for Urban Studies of the University of North Carolina. Seeking to fulfill the need for information about the interrelationships between American lifestyles and energy use, participants discussed the best ways to adapt to changes in energy supplies and prices. This volume draws together reports of recently completed and on-go-ing research on the relationship of energy to American communities and households, assessments of how Americans can best establish more efficient use of energy resources, and detailed outlines of current research needs. Bringing about such a transition with a minimum of economic and societal disruption is the major challenge for urban planners, policy makers and researchers, *ENERGY AND THE COMMUNITY* provides a concrete base to meet this challenge.

Energy and the Community

Edited by
Raymond J. Burby, III and A. Fleming Bell

Center for Urban and Regional Studies
The University of North Carolina at Chapel Hill

Ballinger Publishing Company • Cambridge, Mass.
A Subsidiary of Harper & Row, Publishers, Inc.

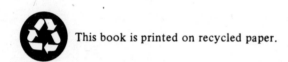
Copyright © 1978 by Ballinger Publishing Company. All rights reserved. No part of this publication may be reproduced, stored in a retrieval system, or transmitted in any form or by any means, electronic mechanical photocopy, recording or otherwise, without the prior written consent of the publisher.

International Standard Book Number: 0-88410-083-9

Library of Congress Catalog Card Number: 78-15760

Printed in the United States of America

Library of Congress Cataloging in Publication Data

Colloquium on Energy and Patterns of Human Settlement, University of North Carolina at Chapel Hill, 1977. Energy and the community.

    1. Energy consumption—United States—Congresses.   2. Energy conservation—United States—Congresses.   3. Community development—United States—Congresses.
I. Burby, Raymond J., 1942-   II. Bell, A. Fleming.   III. Carroll, T. Owen.
IV. Title.

HD9502.U52C647 1977   333.7   78-15760
ISBN 0-88410-083-9

Contents

List of Figures

List of Tables

Preface

The initial conception of this volume arose out of a concern with the implications of the energy crisis for our communities. Where we live and the ways in which we live have been shaped by the unprecedented personal mobility associated with the automobile and an abundance of low-cost fossil fuels. As these fuels become more precious and energy costs rise, patterns of living and settlement must change. Bringing about the transition to more energy-efficient communities with minimum disruption of present life-styles is a major challenge facing urban planners, policy makers, and researchers.

As a first step in addressing this national problem, we undertook a critical examination of the various avenues for local government intervention to affect energy conservation. The results of this exercise were published in late 1976 in *Energy Conservation: A New Function for Local Governments?* by Winston Harrington. Although it was found that local governments may have an important role to play in energy conservation and development, it was also apparent that a number of serious research gaps must be closed before appropriate policies can be delineated. Fortunately, much of the necessary research was already under way. Thus, it was a logical next step to convene a Colloquium on Energy and Patterns of Human Settlement on the campus of The University of North Carolina at Chapel Hill during the spring of 1977. The articles in this volume emerged from the six colloquium sessions that were conducted to explore current research results, fruitful methodological approaches, and priorities for the next round of research on energy and the community.

We owe a great debt to a number of people who helped us identify important lines of research to be included in the colloquium. In particular, we are grateful to Joel Darmstadter and Winston Harrington of Resources for the Future; Corbin Harwood of the Environmental Law Institute; Kim Gillan and Robert Stephenson of the U.S. Energy Research and Development Administration; Hugh Knox of the Federal Energy Administration; and Roger Hansen of the U.S. Environmental Protection Agency. On our campus at Chapel Hill we would like to thank the members of our advisory committee, Jonathan Howes, Susan Clarke, and Daniel Textoris, for their help. We received additional advice from Thomas Donnelly, Gorman Gilbert, Maynard Hufschmidt, Edward Kaiser, David Moreau, Barry Moriarty, Philip Singer, and Joseph Straley. We would like to acknowledge the contributions of Barbara Rodgers, who transcribed the colloquium sessions and typed the final manuscript; Jennifer Jenkins and Deborah Wahl, who typed the drafts and Carroll Carrozza, who supervised production of the manuscript.

A final note of thanks is due the fourteen speakers who participated in the colloquium and whose research is summarized in this volume; our colleagues from throughout the university community who took part in the colloquium sessions and whose incisive comments added greatly to the success of the endeavor; and to the university's Council on Urban Studies, for the grants that have made possible our research on energy and the community.

Raymond J. Burby, III
A. Fleming Bell

Chapel Hill, North Carolina
November 1977

Introduction

It is clear that energy shortages and rising energy prices are becoming a way of life in America. The oil embargo of 1973 and 1974 and the natural gas crisis of 1976 can no longer be seen as isolated incidents; they are but the first signs of the problems that Americans, the greatest consumers of energy in the world, will experience in the next few years as we attempt to change our present patterns of energy use.

"Conservation" is touted by many, including the present federal administration, as one of the keys to a brighter energy future for America. We are urged from all sides to change our personal habits, our methods of conducting business, and our ways of building homes and cities, so that energy, and perhaps the country, can be saved. The average person listening to such cries finds himself or herself in somewhat of a quandary. Many of the voices heard are contradictory. Some speakers are self-serving, and others may not be fully informed. Plainly put, we do not know enough about how we can best become an "energy efficient nation." One doubts whether we can even agree on a definition of the phrase.

Some examples will help to illustrate the point. We are not sure whether it is more energy efficient to build compact, high-density housing, or to construct sprawling, low-density units with banks of solar collectors. We do not know whether it is a better public policy to provide transit service to speed people from the outlying parts of metropolitan areas to the central city, or to bring more stores and jobs out to the suburbs so that people can make short, efficient trips by personal automobile. We do not know enough about building reasonably priced homes and offices that use as little energy as possible to provide a given level of comfort. We do not know whether it is better to build nuclear power plants in rural areas with minimal population at risk but sometimes severe socioeconomic disruption, or to locate power plants near the cities served, where the population threatened by a mishap is greater but the socioeconomic impact is much smaller. In short, whereas we know that we do not want to run out of energy, we do not know how to arrange or operate our communities to most efficiently meet our energy needs.

Recognizing the lack of information about the interrelationships between energy use and American life-styles, the Center for Urban and Regional Studies of The University of North Carolina at Chapel Hill convened a colloquium on Energy and Patterns of Human Settlement in the spring of 1977. The Center brought together many of the outstanding researchers in this field to discuss research findings and priorities and to exchange ideas about the best ways for Americans to adapt to changes in energy supplies and prices.

Energy and the Community presents the results of this effort. It draws together reports of recently completed and ongoing research about the relationship of energy to American communities and households, thought-provoking assessments of how Americans can best go about establishing more energy efficient life-styles, and detailed discussions of current research needs.

In Part I of the volume, People, Buildings, and Energy, Eunice S. Grier and Stephen D. Julias, III set the stage for a consideration of community energy use with a discussion of recent empirical research on how Americans use energy in their homes and for transportation. Both writers agree that income is a major factor in determining energy consumption levels, with higher income families accounting for a proportionately greater share of total energy use. Each argues that increases in fuel prices to curb consumption will have little effect on the rich, while unduly hurting the poor, who already spend a far greater part of their incomes on energy than do those who are better off. Grier notes that a large amount of energy waste occurs at all income and consumption levels.

David T. Harrje and Michael Sizemore both suggest that much energy inefficiency or waste is a result of the ways Americans construct and operate buildings. Harrje reports achieving substantial reductions in energy consumption in townhouses in Twin Rivers, New Jersey, by sealing cracks, adding insulation, and otherwise modifying poor construction. Sizemore estimates that energy consumption in existing and new buildings could be cut by 30 to 60 percent by means of simple retrofitting and changes in operating procedures and construction practices.

Large numbers of the individual consumption decisions discussed by Grier, Julias, Harrje, and Sizemore combine to form the complex energy use patterns of a metropolitan area. In Part II, Land Use, Transportation, and Energy, James S. Roberts reports on a pioneering study of the Washington, D.C. area, which was one of the first research projects to document the relatively high energy consumption caused by sprawling, low-density development. Roberts also discusses his current efforts to compile and evaluate the available data on the interrelationship of energy use and land use patterns.

Robert's early work is expanded by T. Owen Carroll in a study of the energy implications of two alternative development patterns for the Long Island, New York area. Carroll reports that total 1992 energy demands for Long Island, calculated using very sophisticated energy-intensity factors, should be much lower with a compact pattern of development around urban centers than with urban sprawl. He notes that a workbook has been developed to aid community planners in determining the energy implications of alternative land use patterns.

Jerry L. Edwards is concerned primarily with the influence of land use on the energy needed for transportation. After testing thirty-seven hypothetical urban development patterns, Edwards concludes that compact cities use less transportation energy but are usually less accessible than cities with sprawling development. Recent research by Edwards and others indicates that polynucleated cities, with a number of small urban centers instead of one large downtown area, may perhaps provide the best balance between energy efficiency and accessibility.

Although changes in individual behavior, in building practices, and in land use patterns to achieve energy conservation look fine on paper, Dale L. Keyes, Grant P. Thompson, and Bruce Hannon all suggest in Part III, Community Energy Conservation in Perspective, that they may be very difficult to implement. Keyes finds that the quantity of energy savings that would be provided by 1985 with a large but achievable shift in land use patterns would be less than 0.35 percent of total United States energy consumption. He suggests that government should act as a facilitator, by means of tax incentives or other means, of energy conservation by the private sector, rather than as a strict regulator of what types of development take place.

Thompson largely agrees with Keyes's view. He notes that it will be extremely difficult to make someone build a house so that it saves energy, if there is no economic incentive to do so. Thompson suggests, however, that it is probably inevitable that government will increasingly use building codes to regulate the energy efficiency of development.

According to Hannon, higher energy prices are the only certain way to bring about energy conservation. Hannon proposes that a tax be levied on energy at its source, in order to prevent windfall profits by energy companies and to spread energy price increases equitably over all goods in the economy. Following a detailed discussion of past and present energy use in the United States, he explains the futility of trying to offset increasing energy costs by working longer hours or by adopting an energy-conserving life-style.

Although energy is usually thought of as a dependent variable to be manipulated by individuals and groups in the community, John H. Cumberland and John S. Gilmore show convincingly that energy sometimes assumes the upper hand. In Part IV, Energy Production and the Community, each discusses the severe socioeconomic problems that development of new energy supplies has caused for small, rural communities in Maryland and Wyoming. Declines in the quality of life and sharp increases in the costs of providing government services are two of the major problems cited by both Cumberland and Gilmore. Each asserts that the increased tax revenues which energy development projects provide do not become available until long after the damage is done.

Stephen S. Skjei takes a different view of the problems of energy facility development. Leading off the discussion in Part V, Research Needs in Energy and the Community, he suggests that a considerable amount of additional study is needed before the types and magnitudes of impacts caused by new power plants can be described accurately. He notes that the lack of theoretical and quantitative knowledge in this area severely hampers regulatory agencies in making licensing decisions and evaluating environmental impact statements.

Kim Gillan concludes the volume by discussing other research needs in the area of energy and the community. She explains that even though a number of federally funded research projects are presently under way, little concrete information is yet available on how energy can best be conserved in buildings and localities.

As Skjei's and Gillan's articles indicate, socioeconomic investigation of energy and the community traditionally has been neglected. Most research has focused on the development of new energy technologies, ignoring the fact that new hardware taking many years to refine is hardly a solution to present shortages and price increases. If we are to adapt out lives and our communities to the energy realities of the next ten to twenty years, much additional social and economic research must be conducted. *Energy and the Community* provides a starting point for the needed studies.

✳ *Part I*

People, Buildings, and Energy

Chapter 1

Energy Consumption in American Households

Eunice S. Grier

Most Americans are probably no longer in a state of innocent bliss with regard to the finite limits of our national energy resources or the growing costliness of using them. We surely must have learned a few things from the bitter winter of 1977 and from the Arab oil embargo of 1973–1974, with its long lines at the gasoline pumps. Our growing awareness that the energy supply-demand relationship is changing is reinforced by a recurrent pain in the individual pocketbook, a pain that seems to get a bit sharper every time we fill up at the gas pump or turn up the thermostat to what now seems a barely comfortable level. We are beginning to face the hard reality that it can only get worse before—if ever—it gets better.

From the data on energy use and its correlates that are only now beginning to be amassed, one fact stands out more and more sharply. Our level of energy consumption is very directly and very heavily influenced by our patterns of settlement. In no nation of the world is this connection any more evident than in the United States; yet no other highly developed nation has paid less attention to energy costs in its planning, architecture, and development. Most of our housing and other construction, particularly during the decades of unprecedented growth and prosperity after World War II, seems to have been predicated on the implicit assumption that energy supplies would always remain cheap and plentiful. Many of the resulting land use patterns will be very difficult for us to change. They have literally been set in concrete and steel. Yet, regardless of how long it takes, we must find ways to adapt either land use patterns or ourselves to newly recognized reality. Our options in this respect are only beginning to undergo systematic examination by the design professions, and we are still doing very little in the way of either redesign or retrofitting.

This chapter reviews recent findings about the energy consumption patterns of U.S. households. It is based on data from two national surveys of household energy use by the Washington Center for Metropolitan Studies. The first of these studies was conducted in 1973, at a time when very few Americans yet seemed to comprehend that fossil fuels could not be counted on to last indefinitely or at least as far into the future as most people found it possible to look. That study was supported by the Ford Foundation as part of its energy policy project. Its results have been reported in part in a recently published book, *The American Energy Consumer.*[1] The second survey was supported by the Federal Energy Administration and was conducted in 1975 after the oil embargo had raised energy to the forefront of our national consciousness. In combination, these two surveys provide both preembargo and postembargo measures of household energy use in relation to a wide variety of factors, including socioeconomic levels, housing characteristics, and life-styles.

So far, much of the analysis of the survey data has focused upon the impact of the energy crunch on low-income households. This work has been supported by a grant from the Community Services Administration, a federal agency which has a particular interest in low-income households. The analysis has also involved some comparisons between low-income households and those with high incomes as well as some longitudinal analyses of changes in people's energy consumption habits and their attitudes about energy between 1973 and 1975. Both individually and in combination, these two surveys provide an opportunity to sort out some of the complex interrelationships among household energy consumption, the natural and manmade environments, and life-styles.

SURVEY METHODOLOGY

A brief mention of some aspects of the survey methodology may be useful in interpreting the results. First, both these national surveys used representative samples of U.S. households chosen by area probability sampling techniques. Wherever possible, housing units visited in 1973 were revisited in 1975. In 1975 a completely new set of households, selected along the same lines, was added. As a result, the 1973 sample of about 1,500 households was expanded to 3,200 households. This may seem like a relatively small sample, but it is sufficiently large to identify and analyze both regional variations and differences among various groups in the population. In both years the socioeconomic composition of the samples correspond very closely to the most recent estimates of the Census Bureau's Current Population Survey.

The surveys included a one-hour personal interview with each household. The interviews covered a number of topics. For example, a comprehensive socioeconomic profile of the household was obtained. Other information collected included data about the housing unit itself and its characteristics; data on appliances and energy-using amenities in the home; and data about personal transportation and auto use. A number of questions were asked about life-styles and attitudes toward energy conservation. In 1975, the survey instrument was expanded considerably in the area of transportation, and some items dealing with perceived changes in energy-related behavior and attitudes between 1973 and 1975 were added.

In addition to the household interviews, data on consumption and cost of electricity and natural gas were obtained directly from the household utility suppliers. Virtually all of the families signed waivers permitting the utility companies that supplied their electricity and their natural gas to provide this information. Data were obtained on the amount of electricity and natural gas used by the sample households for each month during the past year, and on how much they paid for it. Fortunately, well over 90 percent of the utilities responded, meaning that the possibility of bias from asking households to estimate their own usage and cost was avoided. These utility data were important, for virtually all American households now have electricity and a very high proportion also use natural gas.

Direct fuel oil consumption data were not collected, although it would have been very desirable to do so. There were just too many fuel oil suppliers within any market area to make data collection feasible given constraints of time and money. However, households in both surveys were asked to estimate their fuel oil costs.

As noted above, much of the analytical work to date has been focused on low-income households. These households are one of the groups that should receive special attention as the nation attempts to develop effective and equitable energy conservation programs. It will require all of our wisdom and human sensitivity to assure that energy programs do not produce further disadvantage for those who are already so vulnerable to disadvantage.

In the analysis, "low-income households" were defined as those households falling below approximately 125 percent of the federal poverty income maxima for the year preceding the survey. These maxima, which are very stringent, also relate income to household size. The federal figures were varied slightly because of the method of collecting data on income. For example, in the 1975 survey, a household of three or four persons was considered to be low income if it received a maximum of $6,000 in 1974. In this case, $6,000 was slightly lower than the federal low-income standard for a four-person household. Roughly one-fifth of U.S. households were low income in 1975, according to our definitions. We defined high-income households in 1975 as those in the top one-tenth of all households; this turned out to be those households with incomes of $25,000 or more.

ENERGY CONSUMPTION AND SETTLEMENT PATTERNS

The data indicate a strong relationship between energy consumption and patterns of human settlement, with the latter defined as both the configuration of dwellings in which people live and the characteristics of those dwellings themselves. A review of several findings illustrates the nature of this relationship.

First, whereas somewhat more than half of all energy used directly by U.S. households is consumed within the dwelling unit itself, an almost equal proportion is used for personal transportation, mainly by automobile. The sprawling, low-density configuration of American dwellings obviously is responsible to a considerable degree for this result.

Within the dwelling, the largest amount of energy by far—57 percent—is used for space heating. This is true despite the multiplicity of appliances used by American households. Space heating requirements are largely designed and built into the dwelling at the time when it is constructed; only minor changes can be made by the occupants after construction. Some of these changes, such as insulating a poorly protected structure, can involve considerable expense. In addition, a poorly insulated dwelling may have other problems that will limit the effectiveness of any insulation.

One key factor in a dwelling's heating energy needs is the degree of exposure of its exterior walls to the elements. On the average, it was found that apartments use only about half as much natural gas as single-family units (Table 1-1). Ths size of the dwelling, which is often closely related to the type of unit, also has a direct relationship to the amount of heat energy consumed.

The geographical location of a home is important in determining its energy consumption. Homes located in the cold north central region of the country use, on the average, substantially more natural

Table 1-1. Average Btu and Average Cost of Natural Gas Consumed Annually by Households in Dwellings of Selected Characteristics, United States, 1975

	Million of Btu per Household	*Average Cost per Household*
All Households	136.3	$224.60
Structure Type		
Single-family homes	147.6	237.70
Apartments	76.3	154.90
Size of Unit		
Under 500 sq. ft.	96.3	161.60
Under 1,000 sq. ft.	110.1	192.70
1,000–1,999 sq. ft.	141.9	225.20
2,000 sq. ft. or more	181.5	292.00
Community Location		
Central city	119.6	197.90
Ring of SMSA	151.2	258.50
Outside of SMSA	138.3	209.60
Region of Country		
Northeast	131.2	281.20
North Central	161.4	235.40
South Atlantic	118.2	234.50
South Central	98.4	146.00
West	121.5	171.00

Source: Washington Center for Metropolitan Studies, National Survey of Household Energy Use, 1975.

gas than those in the south central region. Those in the northeast, which is the second coldest region, likewise use considerably more than southern homes. Heat energy requirements are also correlated with the type of community in which a person settles. Nationally, suburban dwellings use more natural gas on the average than do central city dwellings. Much of this difference relates to variations between central cities and suburbs in types and styles of housing.

Sorting out all of the factors involved in variations in consumption would be an exercise of no practical value. All of these differences are integral parts of our patterns of settlement. Although changes in design can help to make future dwellings less wasteful of energy, most households will have to get along for a considerable while with the dwellings they already have. Everyone cannot be moved from the north central region with its frigid winter to the sun belt states, for example. Existing suburban single-family homes cannot be turned into urban apartment buildings, nor can features which were designed

into many newer homes with little heed to long-term energy consequences be easily changed. Massive glass windows, for example,
which were placed to give a pleasant view of variable nature outdoors
while the occupants are comfortably heated or cooled inside, cannot
be easily moved, despite the fact that they may expose the largest
possible wall area to the prevailing winds. Little can be done about
the fact that the geographic placement of many homes was decided
by the convenience of the freeway network and the prospect of rapid
long-distance commuting rather than by anticipation of the increasing scarcity of gasoline. Like it or not, the nation is stuck with many
of the physical results of its energy myopia. Conservation policies
must be built around these existing structures at least for the immediate future.

ENERGY CONSUMPTION AND
HOUSEHOLD CHARACTERISTICS

Besides illustrating the relationship between energy consumption and
settlement patterns, the data show that energy consumption is very
unevenly distributed in American households. Income level is the
most important single variable affecting consumption. For example,
the 10 percent of U.S. households with the highest incomes consume
in the aggregate almost as much natural gas, one-fifth more electricity, and about twice as much gasoline as do the 20 percent with
the lowest incomes (Table 1-2). A regression analysis of the 1973
energy survey data indicates that income level has far more explana-
tory power over energy consumption than any other variable tested.
    Of course, it is not income alone that determines energy con-
sumption; it is also the material possessions that high incomes permit households to have. High-income households, for example, have
many more cars, and they drive those cars many more miles. Even in
today's auto-oriented America, about half of all low-income households do not have a car at all. The majority of high-income households have at least two cars, and many of them have three or more.
    Higher income families fill their homes with many more energy-
consuming amenities than do low-income households, as Table 1-3
shows. The 1975 survey revealed that 43 percent of the nation's
high-income households have central air conditioning and 32 percent
have window air conditioning. (There is probably some overlap,
for there are households that have both central and window air
conditioning.) In contrast, only 6 percent of low-income households
have central air conditioning, and only 20 percent have window
units. Nearly every high-income household has an automatic washing

Table 1-2. Estimated Aggregate Consumption of Natural Gas and Electricity by Low-Income Households in the United States, Compared to Upper-Middle and High-Income Households, 1975

	Income		
	Low Income	*$14,000– $20,500*	*$25,000 and Above*
Natural Gas			
Total number of households using natural gas (thousands)	9,628	9,416	5,261
Percent of all gas-using households in United States	20%	19%	11%
Aggregate consumption of natural gas (trillions of Btu)	1,057	1,290	1,002
Percent of aggregate gas consumption by all U.S. households	16%	19%	15%
Electricity			
Total number of households using electricity (thousands)	13,937	14,174	7,221
Percent of all electricity-using households in United States	20%	20%	10%
Aggregate consumption of electricity (trillions of Btu)	844	1,580	993
Percent of aggregate electricity consumption by all U.S. households	13%	24%	15%

Source: Washington Center for Metropolitan Studies, National Survey of Household Energy Use, 1975.

machine, compared to less than half of low-income households. About 70 percent of high-income households have automatic dishwashers, but only 5 percent of low-income households own this appliance. Most high-income homes have color televisions, whereas most low-income homes have black-and-white models which consume less energy (Table 1-3).

An examination of the kinds of households in each income class also helps to explain the differential energy consumption by income level. For example, a higher than average proportion of lower income households contains elderly people. Even though a number of these elderly households may have some luxury appliances, possibly acquired in earlier years when their incomes were higher, the data on consumption suggest that they use these appliances very sparingly. Low income households are simply not the kinds of households that merchandisers of high-priced consumer goods usually count as part

Table 1-3. Major Appliances Possessed by Low-Income Households in the United States Compared to Upper-Middle and High-Income Households, 1975 (percent of all households)[a]

	Income		
	Low Income	$14,000– $20,500	$25,000 or more
Central air conditioning	6	23	43
Window air conditioning	20	40	32
Automatic washing machine	47	85	91
Wringer washing machine	15	4	3
Automatic dishwasher	5	35	70
Separate food freezer	26	49	53
Gas clothes dryer	8	25	34
Electric clothes dryer	21	46	51
Gas range or stove	66	49	42
Electric range or stove	33	52	61
Electric refrigerator	98	99	99
(Frost free)	(33)	(67)	(82)
(Requires defrosting)	(64)	(31)	(17)
Gas refrigerator	1	1	1
Black and white television	70	57	63
Color television	35	78	84

[a]Detail will not add to 100 percent because of multiple responses.

Source: Washington Center for Metropolitan Studies, National Survey of Household Energy Use, 1975.

of their market. Far more low-income households than high-income households are headed by women and by blacks, and many more low-income households have an unemployed person at the head.

The kinds and locations of housing occupied by households of various income levels also help to explain differences in energy consumption. Low-income people usually live in small homes. A disproportionate number of them live either in the central cities, in small cities and towns, or on farms. Low-income families do not conform in any way to the image of so-called "typical American households," living in suburban single-family homes with picture windows, well-tended lawns, and two nearly new cars in the garage. More typical of low-income dwellings are cramped urban flats or rundown rural shacks. If there are vehicles at all, they are quite often pickup trucks, vans, or something equally practical, and they are almost certainly a few years old.

If U.S. energy policies for households are to depend heavily upon conservation, it is clear that most of that conservation must take place among middle-income and high-income families. The poor already consume far less than their proportionate share of energy. Use of price mechanisms alone to control consumption will work hardship on lower income households. For example, in both 1973 and 1975 low-income families actually paid more for each unit of energy they consumed than did higher income households, because of pricing policies that rewarded larger consumers by charging them less per unit. Fortunately, the gap in per-unit price paid narrowed somewhat between the two years, and the United States seems to be moving nationally toward a pricing system that charges every residential user within a market area equally at the same per-unit rate. At the same time, many utilities still add a fixed monthly service fee to their per-unit charge. This minimum fee must be paid by the poor, regardless of how little electricity they use.

Many low-income households paid 20 percent or more of their pretax incomes for energy in 1975. This is a much higher proportion of income than that paid by those earning $25,000 or more, despite the fact that the higher income households consumed far more energy. How much lower income households pay for energy is not just of interest to them; the prices they pay affect all of us. Very often these same households are recipients of benefits from such public programs as social security, public housing, and public welfare, whose benefit levels are keyed directly or indirectly to consumer price levels. Although the cost to the taxpaying public of these programs does not increase as rapidly as the living expenses of the families being helped, in the long run we all pay more.

ENERGY CONSERVATION STRATEGIES

A third finding from the analyses is that there is a large amount of unnecessary waste of energy at all income levels. Despite the fact that low-income households consume energy much more sparingly, there is much wastage even among that group. As has already been noted, a considerable amount of this waste is built into the structure of the homes and the structure of the community. Nonetheless, some of the waste could be eliminated if an effort were made to do so. It is paradoxical that many of the same high-income families that had energy-saving features in their homes (63 percent had storm windows, and 74 percent had insulation in both walls and ceilings) reported in 1975 that they had recently made changes, such as adding rooms or changing the wiring to accommodate new air conditioners, that would increase energy use.

There is a great need for public programs to aid the poor in weatherizing their homes. Far fewer low-income than high-income houses were equipped with insulation (24 percent) and storm windows (40 percent) in 1975. Most low-income families cannot afford to install these items without assistance even if they want to do so and the basic quality of the dwelling warrants it. On the other hand, many low-income dwellings may be in such bad condition that weatherization measures are not going to do much good. Perhaps the best thing that could be done for these households is to help them acquire decent housing. This, of course, means more emphasis on publicly assisted housing programs, in nonmetropolitan areas as well as in central cities.

Another practical way to help some low-income families save energy would be to provide them with a means of controlling the amount of energy they consume for space heating. About four out of ten low-income households have no means of controlling the temperature at which they keep their homes. Many of these families rent apartments in buildings in which the heating system as well as the temperature is controlled by the landlord or his agent. The individual tenants have no choice but to accept the level of heating or cooling provided by the central system. If it get too warm, the tenants open the windows. If it gets too cold, they put on more clothes, or perhaps turn on the oven, thus of course also using more energy.

The possibilities for energy conservation in apartment buildings with centralized sources of space heating and single utility meters need to be examined. The data collected on energy consumption and cost cover only the 50 percent of renters who pay their utility bills directly and whose consumption is individually metered and billed. At present there is only scant knowledge of what goes on in apartment buildings where the landlord pays the bills and controls the heat. We do know that at the present time public policies provide few direct incentives either to landlords or to tenants to do more to weatherize buildings or to cut back on energy consumption. Increases in rent when the utility bills go up might sometimes serve as an incentive, but increases are not always directly related in tenants' minds to how warm they keep their homes and how often they turn off the lights.

Even many of the low-income homeowners who live in single-family dwellings have little control over the temperature of their houses. They may have heating units that they can turn on or off but that have no thermostats to allow them to know or to control how warm or how cool they keep their homes. A high proportion of the low-income households surveyed could not estimate their household temperature simply because of this problem. One wonders if it would be practical to add thermostats to the existing heating units in low-income homes. If it were possible to do so, would the cost of modifying these units be justified by the potential savings in energy? In some cases, replacing the existing heaters might make more sense. In other instances, in areas where the climate is relatively mild, simply encouraging families to make a point of turning off the heat whenever the weather permits may be a better alternative.

The automobile market structure has a direct impact on the amount of fuel consumed for personal transportation by low-income households. The pattern of auto ownership, in terms of makes and models, of lower income families resembles very closely that of higher income households. Poorer persons have somewhat fewer luxury cars and subcompacts than those who are wealthier, but in general the ownership patterns of all income groups tend toward the midrange of standard or intermediate models. However, because low-income families generally buy their cars used, they tend to lag behind higher income households in the features found in their cars. This means that the improved gas mileage of those cars being sold now will benefit poorer households only several years from now. For example, even though low-income households had cut back somewhat between 1973 and 1975 on the total number of miles they drove, the gasoline mileage they were receiving had become progressively worse over that period of time. This decrease in mileage occurred because cars built during the late 1960s and early 1970s (the cars poor families were driving in 1975) were equipped with an increasing number of gas-consuming options.

In conclusion, it should be realized that the group called "low-income households," which has been discussed in this chapter, is not a single entity. It includes both elderly people and young families with many children; both renters and homeowners; and both city-dwellers and residents of small towns and rural areas. In short, the characteristics of low-income families are about as diverse as those of higher income households. Thus, any evaluation of lower income households' energy needs, and their capacity to conserve and change their ways, must take these individual differences into account. The same is true of the American population generally.

NOTE

1. Dorothy K. Newman and Dawn Day, *The American Energy Consumer* (Cambridge, Mass.: Ballinger, 1975).

Chapter 2

Conserve or Consume:
How Households and
Neighborhoods Vary

Stephen D. Julias, III

Booz, Allen & Hamilton recently conducted a study for the Federal Energy Administration to learn more about the interaction of land use patterns and energy consumption. This chapter reviews the methodology and major findings of that investigation.

To make the research more manageable, three basic objectives were defined. First, the study sought to increase our understanding of the relationship between residential land use patterns and energy consumption. Improving the existing data base was a related and necessary second objective. Data collection was to be done at a very specific household-by-household level so that the relationship between energy use and household characteristics could be examined. Finally, the study was designed to examine public policy actions in the land use area that could affect the short-term (one to three years) energy consumption pattern.

The first step in conducting the study was selection of a sample of existing metropolitan areas. Because of time and money constraints, the sample was limited to three areas. In selecting the sample, differences in climate, economic base, and transportation patterns (existence or nonexistence of some form of public transit) were all considered, as was the availability of data.

Fairfax County, part of the Washington, D.C. metropolitan area, was selected as an example of an area with moderate climate and without an indsutrial base. The local government was very interested in the study, and the local power company was willing to be very cooperative in supplying data. Chicago was selected as the cold climate city. It is an industrial city with a rapid rail transit and a fairly well-developed highway network. The third city selected was Tucson. Besides having a warm climate, there is low use of public transportation in this fairly dispersed, rapidly growing city.

Four different types of neighborhoods were examined in each metropolitan area. As nearly as possible, the study was designed to compare areas of high-density housing with low-density areas. Other factors considered in the selection of neighborhoods were distance from major commercial centers, access to various modes of transportation, and diversity of housing types within each sample area. Within each neighborhood only homes using metered natural gas or electricity for heating and cooling were included in the research, since data were not available concerning home fuel oil deliveries.

The next step in the research was to obtain energy data. On the first "pass" through most utility companies, many reasons were given why the information that was needed for the study could not be made available. However, once questioning was pursued to a senior enough level, every utility became willing to supply the needed household consumption figures.

Two major difficulties were encountered in using the utility data. First, most utilities kept track of individual home meter readings on the basis of each customer's name. This meant that if a customer had moved into a home during the analysis period, it was necessary to find out from whom he or she bought the house in order to obtain a continuous record. The other real problem with the data was that in some areas common metering of multifamily units was widespread. In these cases there was no way of determining the energy consumption of individual households.

To make the energy data useful, they were paired with detailed information on each individual or household. Telephone interviews with households were conducted by graduate students from the State University of New York at Stony Brook. The interviews were used to obtain details on family size, family income, ages and number of family members, size of housing unit, type of appliances, type of heating fuel used, and travel habits. Some questions were very difficult for the families to answer. For example, many people did not know how many miles they drove each year. Realizing this, supplementary questions, such as the distance from the person's home to their place of work and major shopping center, were asked so that the research team could reconstruct the necessary information.

As one of the last steps in data collection, a literature search was performed in order to find out what had been done previously in relation to energy consumption. Extensive interviews were also conducted with knowledgeable individuals, as a method of exploring policy options. In particular, the interviews were designed to determine what types of land use controls would really work, and what the impacts of those controls would be, both politically and economically.

The analysis of the data was straightforward. In a series of regression analyses, income, housing size, housing structure, and similar variables were compared to energy consumption. At the same time local governments were asked what types of land use controls could be instituted to achieve the assumed greater efficiencies of high-density living patterns and shorter trips to work.

The first finding from the regression analyses was that a large variance, as high as 25 percent, in energy consumption existed between essentially identical housing units in the same neighborhood. This result immediately called into question most previous assumptions regarding land use and energy. Two other less startling findings emerged from the study. Dwelling unit size was the variable most closely correlated with energy consumption; the larger the house, the more energy was consumed. A meaningful correlation between income and the amount of energy consumed, both within the housing unit and in transportation, was also found. For example, a family with an income of $30,000 per year was found to consume twice as much energy for transportation as a family with a $15,000 annual income. It seems that when people make more money, they tend to be sloppier in their energy consumption habits.

This latter finding suggested that certain policies thought to encourage energy conservation may not achieve the intended result. For example, minor increases in fuel costs will only add to the burden on poorer people, who, according to the survey data, are already doing the greatest share of conserving. Such increases will probably not be a sufficient inducement to encourage the real energy wasters—the rich—to drive less or to turn down their thermostats.

A positive correlation between density and per capita energy consumption was found, that is, the greater the dwelling unit density, the larger the amount of energy consumed on a per capita basis. This finding would appear to fly in the face of the idea that high-density living is more energy efficient. On closer examination, however, it can be explained by the fact that in most high-density situations, there tend to be greater numbers of small, one- or two-person households. The actual space "consumed" by each individual in a smaller household is greater than is the case in a household with more people. This reinforces the finding noted previously that the more space one has, the more energy one consumes.[1]

NOTE

1. For a more detailed account of the results of this study, see Development Research Division, Booz, Allen & Hamilton, Inc., *Interaction of Land Use Patterns and Residential Energy Conservation* (Bethesda, Md.: Booz, Allen & Hamilton, Inc., FEA Task Order CO-04-50250-00, Job 13392-005-001, October 20, 1976).

Chapter 3

The Twin Rivers Experiments in Home Energy Conservation

David T. Harrje

Over the past four years a research team at the Center for Environmental Studies of Princeton University has been attempting to determine the energy savings that can be achieved in existing housing. Because housing units are replenished at a rate of only 1 to 2 percent a year in the United States, modification of existing units is the most important near-term way of reducing energy consumption in the residential sector, which accounts for about 18 percent of the energy used in this country.

The field laboratory for these studies is the planned community of Twin Rivers, New Jersey, located about fifteen miles from the Princeton University campus. The development contains a central shopping area, surrounded by large groupings of townhouses and condominia, with an outside rim of single-family detached homes. On the extreme outer edge of the community is an industrial belt.

The townhouses, which form the major part of the community, are essentially identical. This made it very easy to make unit-to-unit comparisons of energy consumption. It also revealed some very interesting differences in the energy consumption of houses of the same size with the same floorplan. For example, even after controlling for differences in building orientation and other characteristics, it was discovered that some three-bedroom homes consumed twice as much energy as other identical units. These observations indicate that both the behavior of the people living in a home and the structural characteristics of the housing unit are important in accounting for residential energy consumption patterns.

In order to determine exactly how energy was being consumed in the Twin Rivers townhouses, three houses were outfitted with a sophisticated instrument package that monitored every window, door, and appliance, as well as numerous temperatures and air flows, and that recorded the information every twenty minutes. The package included a separate meter on every electrical appliance, and miniature temperature probes above every door in every room to measure changes in household temperature. Small propellers were placed in all of the heating and cooling ducts to measure the flow of heating or cooling air. In order to determine how the people living in the houses influenced energy use, sensors were placed on the thermostats to indicate how often they were adjusted, and switches were used on all outside doors and windows to count how many times and for how long they were opened.

A broader sample of 30 townhouses was outfitted with a smaller package of equipment. Measurements were made of temperatures at every level in the house, use of the major appliances and the furnace or the air conditioner, the number of door openings, the total electricity consumption, and air infiltration. The instrument package used to collect this information was mounted on a panel above the clothes washer. Data were recorded on a special cassette that was removed only once a month, which made it possible to gather information from a number of widely scattered houses simultaneously.

Models were also used to gather information; for example, wind tunnels and models of houses were used to measure air infiltration and air flow. Air infiltration is a serious problem in energy conservation, for about one-third of the total energy in a house is lost as air flows into the house, is heated or cooled and moves out of the house. The wind tunnel helped to determine how this air moves through and around dwellings.

Buoyancy effects, or hot low-density air leaving through the attic, were not overlooked. To study this aspect of air movement, tracer gas techniques were used. A special instrument released a harmless gas inside a house and the gas was pumped through the house's duct system by the furnace or air conditioner blower fan. Samples of the air were taken every fifteen minutes, and gas chromatography techniques were used to measure changes in the concentration of the tracer gas.

The studies of air infiltration showed a direct relationship between the wind velocity and the air infiltration rate. A rate of one air exchange per hour was common when there was a moderate wind. If, however, the wind was high (20 mph) and slight window openings existed, the rate grew to two or even three air exchanges an hour. Open fireplace dampers also raised infiltration rates. Infiltration commonly was enhanced by the shafts between basement and attic (such as the plumbing shaft or the shaft around the chimney). According to a recent study, even the wall receptacles may be responsible for a sizable percentage of the air coming into houses.

The Twin Rivers experiments have indicated that a number of factors must be considered in determining how energy is used in a house. "Free heat" is an example of an often overlooked variable that must be taken into account. Free heat comes from the sun, from the appliances, and to a lesser extent, from the house occupants. Free heat sources are like satellites to the main heat source, the furnace. For example, when the temperature was 0 degrees Fahrenheit, the furnace supplied more than 80 percent of a townhouse's heat. If instead, however, the temperature was near 50 degrees, less than half of the required heat to make up for losses had to be supplied by the furnace. The rest was furnished by the free heat sources. How much time the furnace worked really depended in a linear way on the outside temperature conditions and on how much free heat was available. A tight well-insulated house may not require any furnace heating until outside temperatures in the lower 50s have been reached.

The research in Twin Rivers indicated that 25 percent of a townhouse's gas furnace heat went up the flue, another 25 percent went to the basement, and the rest went to heat the rooms. Of this latter 50 percent, however, only part heated the rooms directly. Because the ducts passed through the walls in the first floor to get to the second floor, the walls were heated before the second floor rooms, often resulting in localized heating problems in some rooms.

In the summer, the processes were quite different. The inside air was cool and dense because of air conditioning, and it sank to the lower levels of the houses. In contrast with the heating season, in summer only 7 percent of the heat gain occurred by air infiltration. During this season, the dominant factor was the amount of direct energy from the sun allowed to enter the home. The shading of windows in the summer is thus a very important way to keep down the air conditioning load. Limiting appliance usage can also help. Making use of the day-night temperature cycle becomes attractive in the well-insulated home. By opening and closing doors and windows correctly, one can take advantage of natural air movement and the outside temperature cycle.

Infrared photography was used to determine exactly where heat loss was occurring in the Twin Rivers townhouses. Pictures from an infrared scanning camera indicate temperature differences within a house by differences in color. For example, a cold wall with uninsulated areas appears blue in a photograph, whereas a well-insulated wall is yellow. Gradations of temperature are indicated by gradations of color from blue to yellow.

Infrared photography disclosed, as noted previously, that the heat ducts going to the upstairs rooms in the townhouses lost much of their heat downstairs. It was learned that much energy was lost through the fire walls between units, through ducts improperly placed near outside walls, and through improperly installed insulation. Attics and certain parts of basements also contributed heavily to heat loss.

These very detailed infrared pictures of key energy loss areas in the townhouses were used along with the other data collected to plan the retrofitting of some of the dwellings. The objective in retrofitting was to use the least money to get the greatest result in the shortest period of time. In order to get more of the heat from the furnaces into the houses, ducts were insulated and unneeded openings around the furnace flues were closed. (These openings were allowing warm air to move up to the attic and out through the attic vent.) Because the infrared photographs indicated that a considerable amount of heat was lost to the attics, as much as another R–19 of cellulose or fiberglass insulation was added to the insulation already in place, and cracks in attic areas were sealed. These efforts in the townhouses' attics resulted in total annual savings in heating and cooling costs of more than 15 percent. Insulation was also placed wherever spaces existed between masonry and woodwork, and seals were placed on windows and doors. These steps reduced the houses' infiltration rates to 60 percent of what they were originally. One example of rapid payback was the water heater. By wrapping an electrically heated unit with two inches of fiberglass insulation at a per-unit cost of about $5.00, savings of approximately $3.00 per month were achieved. Lower water storage temperature saved more energy. The simple step of cutting back house thermostats resulted in major savings. These savings were approximately 5 percent per degree Fahrenheit on a twenty-four-hour basis, and 1.5 percent per degree Fahrenheit for eight-hour night setback. Setbacks at the time of the Arab oil crisis resulted in immediate 12 percent reductions in heating bills in the community. The retrofits saved another 25 percent of winter heating bills.

It should be noted in closing that the retrofitting done at Twin Rivers increased comfort at the same time that it reduced operating costs. Uniform room temperatures and an absence of drafts and cold spots in walls were common features of the retrofitted townhouses.[1] This increased comfort allowed one to move to other energy saving approaches, such as night temperature setback.

NOTE

1. For a more detailed account of the early results of Princeton's "Energy Conservation in Housing" project, see David T. Harrje, *Retrofitting: Plan, Action, and Early Results Using the Townhouses at Twin Rivers*, Report No. 29 (Princeton, N.J.: Center for Environmental Studies, June, 1976). Further details appear in a book published by Ballinger under the title *Saving Energy in the Home, Princeton's Experiment at Twin Rivers*, edited by R. H. Socolow.

Chapter 4

Saving Energy in Buildings

Michael Sizemore

As experience has been gained in architectural energy planning, the need to consider the interrelationships among a large number of variables to conserve energy in building has become more and more apparent. For example, when an energy analysis of an existing building is performed to reduce energy costs, all the factors in the building are so intimately tied together and interrelated that if one wrong calculation is made at the beginning of the study, the entire analysis may have to be repeated. In many instances the interrelationships among factors influencing energy use are more important than the individual items that are being examined. In fact, the term "energy conservation" is somewhat misleading; "energy tradeoffs" would be more accurate. In general, if energy is saved by a course of action, other resources or other ways of life are also affected, perhaps in negative ways. If a university decides to turn off all the lights on campus at ten o'clock in order to save electricity, for example, the janitors who are on duty will find that they have more work to do. Our concern in buildings—and, more recently, in small communities—is with finding the most effective and the most sensitive elements to alter within this web of interrelationships.

There is an enormous potential for energy conservation in buildings. If one could reduce the energy consumption in future buildings by 60 percent, the amount of energy saved would be as large as the amount of energy expected to be derived from nuclear power in 1990, and almost as great as the expected contribution from coal in that year. This percentage is a realizable goal. Perhaps now that we have a president who is advocating energy conservation, we will begin as a nation to try to achieve such a goal.

Unfortunately, however, many attempts to save energy in buildings result in a waste of money, because the cost-effectiveness of alternative strategies is not considered. An elementary school in Atlanta illustrates this point. The school was built with the most advanced, state-of-the-art solar heating and cooling equipment available. The solar equipment cost $650,000, probably about as much as the school itself. However, the windows in the school were not caulked well and leaked huge quantities of air. The doors were mounted so poorly that they would not close properly, meaning that breezes could continually blow through the doorways. In addition, the light switches were installed in such a way that the lights in unused rooms could not be turned out. For $25,000, the energy load in this building probably could have been cut 25 percent; in addition, the cost of the solar system could have been reduced greatly.

In order to determine the most effective ways to save energy in a particular building, a cost-to-savings ratio can be calculated for each proposed modification. Such a ratio compares the investment necessary to save a unit of energy to the rate at which energy savings will be achieved. For example, if one spends $5,000 for storm windows for a dormitory, and the windows reduce energy costs by $1,000 per year, the payback period for the investment is five years. If an alternative energy-saving scheme has a payback period of fifteen years, the storm window investment is a better choice. Similarly, it makes sense to make inexpensive modifications of operating procedures to see if they achieve the desired results before proceeding to more expensive strategies, such as solar space heating, which have longer payback periods. A 20 to 40 percent savings with immediate payback is often possible by operational changes.

The payback rate for an energy-saving modification will vary according to the amount of energy presently being used that will be affected. A hospital, which uses a considerable amount of hot water, for example, will recover its investment in a solar hot water heater faster than will a school or office building. A "law of diminishing returns" applies to some energy-saving techniques, such as the installation of insulation. Beyond a certain point, as insulation thickness is increased, the amount of added savings produced by each additional inch is reduced.

Failure to consider the respective paybacks of alternative energy-conserving strategies is fairly common. As much as $200,000 could have been saved in the cost of the solar system for one particular project, if modifications in the design of the building had been planned along with the installation of solar equipment. In addition, the methods of calculating the loads on the systems were not appropriate for solar systems. The correct techniques for a solar heating system are drastically different from those conventionally used.

Data on the energy consumption of similar buildings in different parts of the country show the importance of building operation in accounting for energy costs. Many buildings in Atlanta, a city with about 3,000 degree days per year, use as much energy as similar structures in Princeton, which has about 5,000 degree days (24-hour periods during which the average temperature is one degree below 65 degrees Fahrenheit). The same sorts of variations have been noted between similar buildings in the same city. Although the differences in the air conditioning loads of the Georgia and New Jersey climates account for some of the variation, more important are the ways in which the respective buildings are operated. Some building owners heat and cool their buildings all night and on weekends whereas others have evening setbacks of thermostats. Some owners maintain their heating and cooling equipment better than do others.

Often, very simple modifications in operating procedures can result in large savings in energy costs. At one university, for example, an indoor swimming pool used as much energy as the multistory classroom building located next to it. The heated pool water gave off hot vapor because the outside air was cold. To avoid mildew from the accumulated vapor, enormous amounts of cold outside air had to be brought in at night and heated to keep the air dry. By simply covering the pool with $200 worth of plastic at night, the utility cost was reduced by $15,000 per year.

The habits and dress of building occupants also make a difference in energy consumption. For example, in a bank in New Mexico, all the women wore sleeveless uniforms and all the men wore wool suits. The women wanted the air conditioner set at one temperature while the men wanted it set at another, and the two groups were constantly fighting each other, as were their respective mechanical systems.

Unfortunately, new working and living spaces are often designed with neither comfort nor conservation in mind. Current office building design is an example. Some buildings have large expanses of glass and metal through which the sun's heat can penetrate very quickly. The people working in the offices feel very hot because of the superheated windows, even though the thermostats indicate that overall room temperatures are comfortable. Had these buildings been built with concrete or masonry panels instead of metal, they would not heat up as quickly and the problem would be reduced. Similarly, many patios intended to be pleasant places for people to sit and relax cannot be used during most of the year because of their great amount of reflected heat and their lack of vegetation. Poor design also results in higher than necessary home utility bills. Some homes studied in New Mexico had high energy costs, even though they were located in an area with such a good climate that virtually no heating at all should have been necessary.

Lighting design is a major problem in many buildings. Structures are often provided with an overabundance of artificial lights and are built in such a way that use of natural light is impossible. For example, although it is only four years old and is one of the most efficiently designed and operated buildings in the Washington area, 65 percent of the air conditioning and heating costs in the American Institute of Architects' national headquarters building in Washington, D.C. result from lighting. Projections based on the initial phases of a study of this building indicate that its energy consumption can be cut in half, largely through changes in lighting, with the cost to be recovered within two and one-half years, while concurrently improving building comfort and appearance. Right now the building has the same quality of lighting everywhere. If lighting of individual work areas can be substituted, part of the electricity presently needed for running the lights can be saved. In addition, the tremendous air conditioning load, caused by the heat that the present lighting system is giving off, can be decreased. These changes should make the building a more pleasant place in which to work. Rooms with lighted work areas are generally more agreeable than those with glaring ceiling lights. If the energy consumption of some of the best-designed and best-run buildings can be reduced by one-half, one wonders how much consumption could be reduced if an effort were made to modify both good and bad structures.

Other energy-saving strategies besides those mentioned above are also possible. For example, by designing a new house with ventilators at top and bottom, and by insulating properly, the need for air conditioning can be reduced substantially for very little money. In this case, the design takes advantage of the natural movement of warm air up and out of the house, bringing cooler air in behind it. Another fairly simple way to reduce the air conditioning load is to reduce the amount of paved area in new developments. Many subdivisions have far larger streets than are needed for the amount of traffic to be handled. Researchers who compared temperatures before and after construction on the site of the new town of Columbia, Maryland, found that the addition of a large number of covered surfaces caused a drastic increase in the average temperature of the area.

Use of waste heat is another way to save energy. In one instance, it was proposed that a hospital use the waste heat obtained from burning trash to heat hot water. In this case, the necessary heat recovery equipment could be paid for in less than six years with the savings from reduced fuel bills. Energy use can also be reduced by designing buildings with mechanical equipment and insulation adequate for the normal seasonal heating and cooling loads. This seems more sensible than following the common practice of "building in" extra capacity in anticipation of peak demands that rarely occur.

As time goes on, people developing energy-saving strategies such as those mentioned will be forced to pay increasing attention to the energy used to produce the energy-saving equipment that they plan to install. For example, one particular solar collector will have to be used for three years to collect enough energy to replace the energy used to produce the aluminum in that same collector.

Unfortunately, much of the information needed to make the sorts of calculations concerning energy costs and paybacks that have been described in this article is not yet readily available to the individual homeowner and building operator. However, the states are beginning to produce some materials, and the American Institute of Architects (AIA) has recently published an *Energy Notebook*, which has a monthly newsletter and a quarterly update. The AIA publication contains very good technical information and a case study about energy and the built environment.

The material being published today can be very helpful. However, it must be remembered that successful energy planning is difficult. It requires the manipulation of a very large number of variables, ranging from cultural attitudes to mechanical equipment. In the end, if energy is to be saved in buildings comprehensively and efficiently, it will depend on the professional's ability to understand and work with these variables.

✳ *Part II*

Land Use, Transportation, and Energy

Chapter 5

Energy Conservation and Land Use: Prospects and Procedures

James S. Roberts

The relationship between energy and land use patterns is as important as the use of energy in individual structures.

Unfortunately, it is often overlooked. Because of the patterns of urban development that have evolved over past years, a built-in component of the demand for energy has been created. Because of the location of particular structures housing a number of urban activities, transportation and infrastructure networks have developed that assume a certain available level of energy. Preferences expressed for housing, employment, shopping, and social, recreational, and educational opportunities imply particular levels of energy consumption.

There is another dimension to the relationship of energy and land use. Because of presently rising energy prices and future energy shortages, there will be alterations to urban form and land use patterns. The design and location of urban activities will have to take account of changes in both the price and availability of increasingly scarce energy supplies.

This chapter summarizes the prospects for reducing energy demands by systematically altering urban form and land use patterns. Such changes will occur over a relatively long period, and will be exacted at some cost to consumer and other preferences. Because land use changes may successfully be guided and shaped only by the activities of regional and local governments—according to the traditional and well-established role of government decision-making in shaping growth—there is a need to explore how planning, analysis, and implementation of energy conservation by means of regional and local land use can be accomplished. A second purpose of this chapter is to outline briefly a methodology or set of procedures useful for devising an energy conservation program oriented to the needs and requirements of regional and local governments.

ENERGY REDUCTION PROSPECTS

In order to understand the relationship between urban growth patterns and energy consumption, it will be useful to refer to a research effort undertaken several years ago for the Metropolitan Washington Council of Governments.[1] As part of an overall examination of growth policies in the Washington, D.C. area, a study relating energy, land use and urban form, and growth potential for the metropolitan area was undertaken.

The first step in the study was to determine the pattern of energy consumption in the Washington metropolitan area for the base year of 1973. It was found that residential uses accounted for 31 percent of total consumption, whereas transportation represented 29 percent. Commercial, industrial, and institutional uses were responsible for only 40 percent of the area's energy consumption, reflecting the fact that Washington has little manufacturing or industry.

Next, a set of six "alternative futures" or scenarios for the location of the new development predicted for the metropolitan area through the year 1992 was constructed. These alternative arrangements of the 500,000 new households and 780,000 new jobs which Washington is expected to have added by 1992 included the following categories.

1. *Wedges and Corridors*
   Development is assumed to occur in the way mandated by the Council's comprehensive plan for the year 2000, with growth concentrated along a number of transportation radials and with open space preserved inside the "wedges" created by these "corridors" of development.
2. *Dense Center*
   The increment of households and jobs will all be allocated in close proximity to the metropolitan center, largely in the District of Columbia, Alexandria, and Arlington.
3. *Transit-Oriented*
   High-density development of employment opportunities and housing occur near stops on the Metro subway system, which is assumed to be completed by 1992.
4. *Wedges and Corridors with Income Balance*
   The household and employment allocations derived under the regular "Wedges and Corridors" plan are maintained,

but differences in average household income in different parts of the metropolitan area are reduced. This scenario was used to test the effect on vehicle miles of travel of placing higher income households—which take longer, more frequent automobile trips—in all parts of the metropolitan area.

5. *Sprawl*

Residential development will occur at low densities on the urban fringe, whereas new employment will be concentrated in the metropolitan center and new retail development will be located near the Capital Beltway, which circles the metropolitan area.

6. *Beltway-Oriented*

All new development is to be located on vacant land adjacent to the Capital Beltway.

2054229

In order to estimate residential energy consumption under each scenario, per-unit household energy consumption factors for each housing type were multiplied by the number and density of housing units to be located in different parts of the metropolitan area. It should be noted that the consumption factors available at the time the study was done were not very detailed, and that better measures are being developed. Because of the great amount of variation of consumption within the commercial sector and the consequent lack of reliable consumption factors, energy consumption in that sector was held constant under all scenarios, and was included only for bookkeeping purposes.

Given the allocations of households and employment that had been made, a transportation model was used to determine the number of vehicle miles traveled under each scenario. Sprawl had the largest number of annual vehicle miles, followed by Wedges and Corridors. Transit-Oriented had the lowest number of total vehicle miles traveled. Because the Metro system was assumed to be the same under all alternatives, public transit was held constant.

The estimates of energy consumption by alternative scenario are presented in Table 5-1. As can be seen from the table, the Sprawl scenario consumed the most energy, whereas the Dense Center and Transit-Oriented schemes consumed the least. It can also be noted that the Wedges and Corridors with Income Balance scenario consumed less than the regular Wedges and Corridors plan. Even though higher income people were assumed to use their cars no matter where they lived, lower income people who lived in the suburban areas under the income balance plan were assumed to make heavy use of mass transit in commuting into the city for work. This greater use of the more efficient mass transit for long journeys accounted for the lower energy consumption under the Income Balance scenario.

Table 5-1. Energy Consumption by Alternative Development Scenarios (in 10^{12} Btu/yr.)

Consumption by Sector	1973 Base	Wedges and Corridors	Dense Center	Transit-Oriented	Wedges and Corridors with Income Balance	Sprawl	Beltway-Oriented
Residential	265.3						
Increment		109.9	91.0	95.8	109.9	122.6	112.4
Total, forecast year		375.2	356.3	361.1	375.2	387.6	377.7
Commercial/Industrial Institutional	176.6						
Increment		78.9	78.9	78.9	78.9	78.9	78.9
Total, forecast year		255.5	255.5	255.5	255.5	255.5	255.5
Transportation, Automobile	117.9						
Increment		59.5	35.1	33.1	46.8	70.6	52.2
Total, forecast year		177.4	153.0	151.0	164.7	188.5	170.1
Transportation, METRO	2.5						
Increment		12.4	12.4	12.4	12.4	12.4	12.4
Total, forecast year		14.9	14.9	14.9	14.9	14.9	14.9
Total	562.3						
Increment		260.7	217.4	220.2	248.0	384.5	255.9
Total, forecast year		823.0	779.7	782.5	810.3	846.8	818.2

Source: James S. Roberts, *Energy, Land Use, and Growth Policy: Implications for Metropolitan Washington* (Washington, D.C.: Metropolitan Washington Council of Governments, June 1975). Also, see James S. Roberts, "Energy and Land Use: Analysis of Alternative Development Patterns," *Environmental Comment* (September 1975).

In summary, it was found in this study that about 9 percent of the total metropolitan consumption of energy by 1992 could be saved by rearranging land use patterns, whereas 31 percent of the increment in energy consumption could be saved by rearranging. However, these potential reductions in demand should be carefully weighed. There is a terrific price to be paid for rearranging land use patterns; it will not be simple to shift life-styles and people's preferences. Although there is evidence from this analysis that some savings could be achieved by land use means, the disruption that would be caused by such measures could outweigh the advantages.

ENERGY REDUCTION PROCEDURES

Once the relationship between energy and land use is acknowledged, it becomes important to determine how the energy conservation potential inherent in growth and development can be realized. In that regard, it is necessary to summarize a research effort undertaken by the Real Estate Research Corporation, as part of a large-scale multidisciplinary study for the Department of Energy, to develop a set of procedures whereby local planners and other public officials can develop an energy conservation program.

As this methodology for comprehensive community planning for energy conservation and management is envisioned, it will permit the user to develop either a very detailed and analytical assessment of energy use or a less detailed sense of energy consumption, with a focus on potential problem areas. The methodology might be used either to evaluate particular energy conservation measures or to compile a large number of such measures into an implementable energy conservation program. The essential elements of the methodology include the following:

1. A land use, urban activity-based audit of energy supplies and demands in the community.
2. A means of identifying those areas of concern where energy conservation efforts should be concentrated.
3. A set of energy conservation measures with suitable alternative strategies for implementing the conservation measures.
4. A set of procedures for assessing the impacts of the conservation measure and its implementation strategies. Impacts include both energy savings and other effects, particularly socioeconomic changes.
5. Guidelines for the development and implementation of an energy conservation program.

This research effort will be field-tested by a number of local communities. The tools that are developed will be made available to local planners to give appropriate weight to energy conservation efforts in planning for community growth and development.

SUMMARY AND CONCLUSIONS

This chapter has provided a brief description of the prospects for energy conservation by land use means. Although the evidence is preliminary and based on rough estimates, it seems that considerable conservation potential exists. However, it is acknowledged that any land use changes will be difficult to achieve if they are not consistent with prevailing trends. In any case, methodological procedures are being developed that should increase awareness of energy conservation opportunities. These new techniques may in time provide the necessary tools for planning and implementing energy conservation by means of community development.

NOTE

1. James S. Roberts, *Energy, Land Use, and Growth Policy: Implications for Metropolitan Washington* (Washington, D.C.: Metropolitan Washington Council of Governments, June, 1975).

Chapter 6

Calculating Community Energy Demands

T. Owen Carroll

What are the energy implications of alternative land use patterns? How would one calculate the energy consumption associated with each land use sector? These are the questions that a research team at the Institute for Energy Research of the State University of New York at Stony Brook tried to answer during a recent two-year study for the Federal Energy Administration.[1] The Long Island region of New York State served as a case study area for the research. The study was designed to determine how much energy could be saved by changing from Long Island's present urban sprawl development pattern to a pattern of residential clusters and multiple-use centers with substantial open space, as recommended by the Nassau-Suffolk Regional Planning Commission's master plan.

Population and basic employment projections, information on the transportation and services infrastructure, data on the physical characteristics and limitations of the region, and local growth preferences were all used to determine the probable future development pattern in the region under the urban sprawl scenario. The commission's master plan specified growth quantities and locations with the clustered option.

Once the projected quantity and pattern of growth under each scenario was determined, energy intensity, or energy use per unit, factors were determined for different land uses. Factors were developed for different types of residential units and commercial categories. Industries were grouped by energy consumption characteristics into five clases based on the common grouping of industries as light, medium, and heavy. The Standard Industrial Classifications (SICs) often used by planners were not employed, for the SICs frequently place energy intensive industries in the same category with other industries that consume little energy. Transportation energy intensity factors based on a broad review of existing data were developed for automobiles and buses.[2] Subsequently, all energy intensity factors were expressed in a form convenient to application in any region.

Table 6-1 provides an example of the energy intensity information for an average home. The end-use energy demand figures[3] in this table are for a typical three-bedroom home of 1,300 square feet in the northeastern United States. Space heating demands by far the largest share of energy in the home, whereas air conditioning and miscellaneous electric appliances consume relatively small amounts. Heat loss as a result of infiltration through cracks around doors, windows, and basement creates about 50 percent of the space heating demand; direct loss through windows, walls, floors, and ceilings accounts for the other 50 percent. Ceiling heat gain in the summertime is responsible for most of the air conditioning demand. Total energy demand is 135 million Btu annually. This is roughly the same amount of energy as there is in the fuel required to fill the gas tank of an automobile once a week for a year.

Energy demand per degree day[4] is less for higher density residential units than for single-family homes. For instances, in Table 6-2 heating demand is 14,000 Btu per degree day per year in an existing single-family detached unit, but is only 4,300 Btu per degree day per year in an existing high-rise apartment of the same size. The average energy intensity of all types of residential units is less than the average demand of commerical buildings. Residential units require about 10 Btu per square foot per degree day per year, whereas the average annual commercial demand is 15 Btu per square foot per degree day.

The residential, commercial, and industrial energy intensity factors were combined with land use growth projections to produce energy demands by sector for the region. Some of these demands are fuel specific; for example, lighting always requires electricity. For other demands, such as heating, fuel substitutions can be made.

Transportation energy intensity factors were combined with projections of the future numbers of vehicle miles traveled to obtain an estimate of transportation energy demand for the region. National data on the number of daily trips per household for different purposes by different income groups, on mode choice, and on vehicle occupancy rates were combined with local estimates of average trip lengths to determine the number of vehicle miles traveled.

Once sector energy demands and the fuel specific or non-fuel specific nature of those demands were determined for each scenario, a fuel allocation matrix was used to allocate different fuels to the vari-

Table 6-1. Residential Energy Profile

For a Three-Bedroom Home in the U.S. Northeast Region

                1,300 *square feet*
                5,500 *heating degree-days*
                1,200 *cooling degree-days*
                R-7 *ceiling/R-5 walls*[a]

	End Use Demand (million Btu)	Fuel Consumption	Approximate Cost[b]
Space Heat	77 (56%)	102 mcf/850 gal/1,800 kwh	$240/340/450
Central Air-Conditioning[c]	22 (16%)	3,200 kwh	130
Water Heat	18 (15%)	27 mcf/200 gal/5,500 kwh	54/ 80/130
Range	4 (3%)	10 mcf	48
Refrigerator	4 ⎫		
Lighting	3 ⎬ (10%)	4,100 kwh	160
Miscellaneous Electric	7 ⎭		
	135		
	(104 thousand Btu/sq. ft.)		$632 – 918[d]

[a] Weighted average insulation values of existing New England housing stocks as determined for internal FEA report by the Insulation Manufacturers Association.

[b] Natural gas at $2/mcf (thousand cubic feet); oil at 40¢/gal.; electricity at 2.5¢/kwh heat; and 4¢/kwh general use.

[c] Room air conditioning (two units) would be 8.8 million btu.

[d] This range of costs is derived by adding all minimum costs and all maximum costs.

Source: T. Owen Carroll, Robert Nathans, P. F. Palmedo, and R. Stern, *The Planner's Energy Workbook: A User's Manual for Exploring Land Use and Energy Utilization Relationships* (Upton, N.Y.: Policy Analysis Division, National Center for Analyzing Energy Systems, Brookhaven National Laboratory, October 1976), Table 2.

Table 6-2. Heat/Cool Load

Assumptions: 1,300 square feet
15% glass (plus door) area[a]
R-7 ceiling/R-5 wall existing construction[b]
R-11 ceiling/R-7 wall new construction

| | Existing Condition | | | | New Construction | | | |
| | Heat | | Cool | | Heat | | Cool | |
	Btu/°day	Btu/°day per sq. ft.	Btu/°day	Btu/°day per sq. ft.	Btu/°day	Btu/°day per sq. ft.	Btu/°day	Btu/°day per sq. ft.
Single-Family Detached	14000	10.8	22000	17.0	10300	7.9	15500	12.0
Single-Family Attached	10200	7.8	17700	13.6	7400	5.7	12500	9.6
Low Rise	7500	5.8	11200	8.6	5500	4.2	8100	6.2
High Rise	4300	3.3	5700	4.4	4000	3.1	5500	4.2

[a] Percent (glass and door) of total wall area.
[b] Weighted average of insulation values of existing New England housing stock as determined for an internal FEA report by Insulation Manufacturers Association.

Source: T. Owen Carroll, Robert Nathans, P. F. Palmedo, and R. Stern, *The Planner's Energy Workbook: A User's Manual for Exploring Land Use and Energy Utilization Relationships* (Upton, N.Y.: Policy Analysis Division, National Center for Analyzing Energy Systems, Brookhaven National Laboratory, October 1976): Table 4.

ous sectors, with the objective of satisfying demand given certain constraints on supply. The energy demands of the scenarios were then compared.

As Table 6-3 shows, in most cases, the predicted energy demands of the clustered development scheme were less than the requirements of the urban sprawl alternative. The clustered plan was projected to produce only 7.4 percent or 35.7×10^{12} Btu less total energy demand in the year 2000 than the urban sprawl scenario, primarily because a substantial and less energy efficient built environment already exists. Incremental savings in the transportation and residential sectors are large, for example, more than 50 percent reduction in transporation requirements.

The energy intensity factors and other information can be easily adapted to fit any local situation. Residential energy consumption factors, for example, can be adjusted for climate, expressed as the number of heating and cooling degree days, for house size, expressed in square feet, and for the amount of insulation used, expressed as an R value.[5]

In many cases, our data can be used directly. For instance, even though work trip length varies within any metropolitan area, the distribution of work trip lengths according to distance from the city center tends to be the same no matter what size city is being studied. Because work trip length is dependent mainly on distance from the central area and not on population, the figures that have been developed based on national statistics should be applicable in most situations. It should be noted, however, that similarities in trip lengths in different cities do not imply similarities in travel times, unless the transportation networks of the cities being compared are also identical.

A recent publication, *The Planner's Energy Workbook*,[6] contains all information necessary for estimating community energy demands. It provides a step-by-step guide for planners and other local officials who want to consider the energy implications of alternative development patterns for their communities.

Even though it is quite possible to calculate the amount of energy that can be saved by changing development patterns and personal habits, it of course is not easy to convince people to make those changes. Individuals consider many factors besides energy efficiency in making their decisions. An example from the transportation sector illustrates this point. At present, the United States is heavily dependent upon the automobile. Indeed, it has been estimated that automobiles consume roughly 70 percent of the energy used for transportation in this country. Yet, automobiles are highly inefficient users of energy. Why do people persist in using them?

Table 6-3. Summary of Land Use Energy Scenarios for Nassau and Suffolk Counties, New York (Fuel use[a] in 10^{12} Btu)

Sector	Use in 2000 Because of 1972 Population	Incremental Fuel Use in the Year 2000			Difference	Percent of Total Difference
		Urban Sprawl	Clustered	Difference		
Residential	121.1	82.6	70.5	12.1	15	34
Commercial	87.1	63.9	60.8	3.1	5	9
Industrial	28.9	5.5	3.6	1.9	35	5
Transportation	60.5[b]	35.8	17.2	18.6	52	52
Total	297.6	187.8	152.1	35.7	52	100
Total Use in 2000		485.4	449.7	35.7		

[a] Includes electricity at 3,413 Btu per kwh.

[b] Fuel in the year 2000 to provide the number of vehicle miles of travel used in 1972.

Source: T. Owen Carroll, E. Beltrami, A. Kydes, R. Nathans, and P. F. Palmedo, *Land Use and Energy Utilization: Interim Report, BNL 20577* (Springfield, Va.: National Technical Information Service, October 1975): 131.

The major reason for the automobile's popularity lies in the old adage that "time is money." If one multiplies the energy efficiency of various modes of transportation by the average speed of those modes, the rankings of "time-energy efficiency" shown in Table 6–4 are obtained. Although automobiles are less energy efficient than some other modes of transportation, they appear very practical when time as well as energy is considered.

Table 6–4. Time-Energy Efficiency

More Time-Energy Efficient	Large Automobile (general use)
	Small Automobile (general use)
	Commuter Train
	Jet
	Inter-City Train
	Taxi
	Urban Bus
	Automobile (urban use)
	Ocean Liner
Less Time-Energy Efficient	Yacht

As long as other factors are more important than energy in people's decision-making processes, energy conservation and efficiency schemes will not succeed. Only if something can be done to make energy conservation a high priority for most Americans will we have any hope of achieving significant energy savings by means of changes in land use and transportation.

NOTES

1. For an overview of this research, see T. Owen Carroll, et al., *Land Use and Energy Utilization: Final Report*, BNL 50635 (Springfield, Va.: National Technical Information Service, 1977).

2. Reports used included U.S. Bureau of the Census, *Detailed Housing Characteristics 1970* (Washington, D.C.: U.S. Government Printing Office, 1972); Hittman Associates, Inc., *Residential Energy Consumption in Single-Family Housing* (Washington, D.C.: U.S. Government Printing Office, March, 1973); Hittman Associates, Inc., *Residential Energy Consumption in Multi-Family Housing, Data Acquisition* (Washington, D.C.: U.S. Government Printing Office, October, 1972); Federal Energy Administration, *Project Independence, Residential and Commercial Energy Use Patterns 1970-1990* (Washington, D.C.: U.S. Government Printing Office, November, 1974); U.S. Department of Commerce, National Bureau of Standards, *Technical Options for Energy Conservation in Buildings* (Washington, D.C.: U.S. Government Printing Office, July, 1973); and R.L. Knecht and C.W. Bullard, "End Uses of Energy in the U.S. Economy, 1967," Cal Document 145 (Urbana: Center for Advanced Computation, University of Illinois at Urbana-Champaign, 1975).

3. "End use demand" refers to the amount of usable energy required in a house, not to how much fuel is needed to produce that amount of energy. The efficiency of the device using the fuel determines the amount of fuel needed. For example, because oil and gas furnaces are only 60 percent efficient, satisfying a given demand for oil or gas heat takes about 1.6 times as much energy as the amount of final demand.

4. One twenty-four-hour period during which the average temperature is one degree below 65 degrees Fahrenheit constitutes one heating degree day.

5. R values are nationally recognized insulation standards. For example, a house with R-7 ceiling insulation would have insulation about 2 ½ or 3 inches thick.

6. T. Owen Carroll, Robert Nathans, P. F. Palmedo, and R. Stern, *The Planner's Energy Workbook: A User's Manual for Exploring Land Use and Energy Utilization Relationships* (Upton, N.Y.: Policy Analysis Division, National Center for Analyzing Energy Systems, Brookhaven National Laboratory, October, 1976).

Chapter 7

The Effect of Land Use
on Transportation
Energy Consumption

Jerry L. Edwards

During the past four years, research has been under way to examine the impact of the spatial arrangement of land uses on travel behavior and hence on transportation energy use. A spatial interaction computer model, together with data from an existing city, have been used to analyze a number of different spatial arrangements of land uses. From these analyses, conclusions about the sensitivity of transportation energy to changes in urban spatial structure have been developed. These findings have led to other research concerning the energy efficiencies of alternative urban development patterns. This chapter describes these two research endeavors.

The city selected as a data source for the initial research was Sioux Falls, South Dakota. (In 1973 when the work began, data from Sioux Falls were being used by the Federal Highway Administration in much of its work.) Information collected included population and employment data, labor force participation rates (the percentage of residents employed), population-serving ratios, trip rates, auto occupancy rates, and other similar items. As Figure 7–1 shows, these variables were used, together with exogenously specified "media interaction variables" (such as the urban form and the transportation network and levels of service) as inputs to a Lowry-type land use model,[1] which designed hypothetical land use and travel patterns. A transportation gravity model was then used to determine the total amount of travel required for work and nonwork purposes. These travel demands were in turn inputs to energy models for automobiles, buses, and rapid rail transit. The latter models estimated the total daily energy requirement for passenger travel with a given land use pattern.

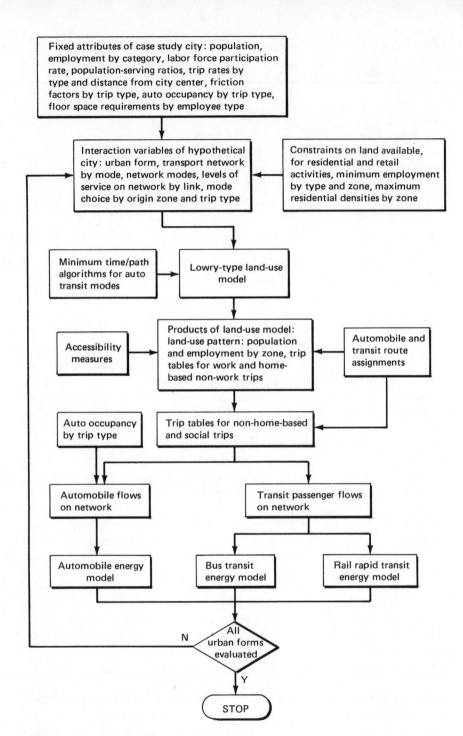

Figure 7-1. Modeling Sequence

To examine the spatial arrangement-transportation tradeoffs of variations of four different urban shapes (shown in Figure 7-2), thirty-seven experiments were conducted. Although the "cities" being studied were hypothetical, the concentric ring and pure linear shapes are fairly common for many American metropolitan areas. Polynucleated cities, with several urban centers instead of one downtown area, are less prevalent in this country. The cruciform city is a variation of the linear and polynucleated shapes.

Figure 7-3 illustrates some of the results of our initial research. Total energy, the vertical axis, refers to all energy used for passenger travel for home-based work trips, for home-based trips to shopping centers and other service establishments, for home-based social trips, and for non-home-based trips. Regional accessibility, the horizontal axis, is a measure of how difficult it is to travel between home and work. The larger the numbers on the respective axes, the greater the quantities of energy used and the greater the accessibility. Each point on the graph represents one of the thirty-seven experimental cities.

All of the cities with high energy consumption (above about 20×10^8 Btu on the vertical axis) were cities where the only mode of transportation operating was the automobile. Those cities using less energy had both public transit and automobiles. The cities with transit were generally less accessible, reflecting the fact that transit passengers pay a tremendous accessibility "penalty" in waiting time and walking time. Cities with sprawling, extensive land use patterns generally tended to score high on both energy use and accessibility, whereas more compact cities usually used less energy and were less accessible.

If those cities in which only automobiles were used are examined, another interesting trend can be discovered. As Figure 7-4 shows, there is almost a linear relationship between total transportation energy used for all kinds of trips, and average automobile work-trip length. This suggests that if one could collect enough representative sample data from an urban area to determine the average work trip length, one might also be able to determine the total amount of energy required for transportation in that city. Unfortunately, the needed data are hard to find. Information is available only on how much fuel comes into a state, not on where the fuel is burned, or on what kinds of trips are made with the fuel.

Cities with both automobiles and transit do not display such a clearcut relationship between total energy consumption and work trip length (Figure 7-5). In these cities, the shape of the urban area acts as an intervening variable, helping to determine the relationship between work trip length and energy consumption. For example, as

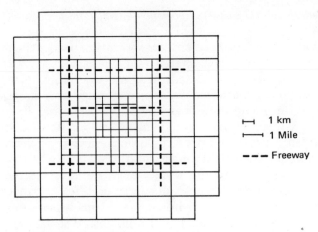

a. Concentric Ring Shape

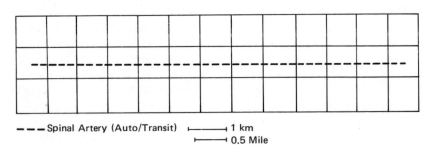

--- Spinal Artery (Auto/Transit) ⊢⎯⎯⎯⎯⊣ 1 km
                                   ⊢⎯⎯⎯⊣ 0.5 Mile

b. Pure Linear Shape

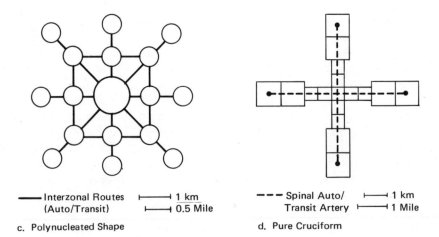

——— Interzonal Routes ⊢⎯⎯⎯⊣ 1 km   --- Spinal Auto/   ⊢⎯⎯⊣ 1 km
    (Auto/Transit)   ⊢⎯⎯⊣ 0.5 Mile   Transit Artery   ⊢⎯⎯⊣ 1 Mile

c. Polynucleated Shape   d. Pure Cruciform

Figure 7-2. Urban Shapes

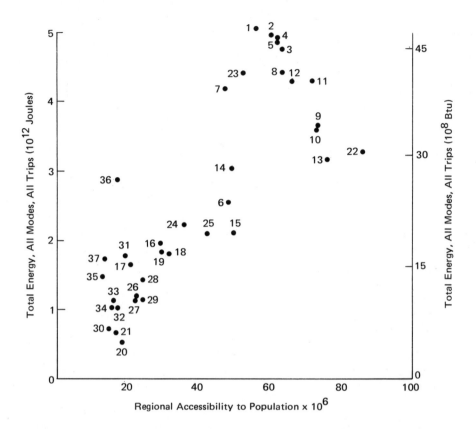

Figure 7-3. Total Energy and Regional Accessibility to Population for Each Experiment

Figure 7-6 illustrates, energy consumption in concentric ring cities rises fairly rapidly with increasing average work trip length, whereas the rate of increase is much lower in polynucleated cities.

This initial research demonstrated that several structural characteristics of cities have important effects on transportation energy consumption and should be considered by researchers. First, four dimensions of urban form must be examined. The (1) shape of the city being studied, whether polynucleated, concentric ring, or linear, needs to be examined, as does (2) the extent to which the city is compact or sprawling (its geographic extent). Population concentration (3) and employment concentration (4) are also important. A city with most employment concentrated in the downtown area,

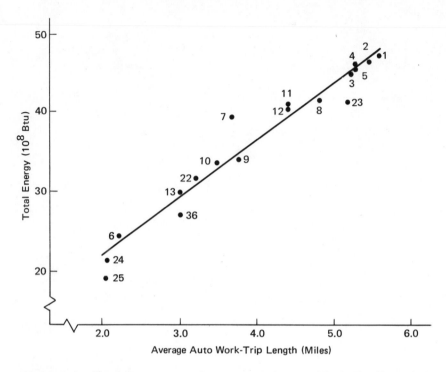

Figure 7-4. Total Energy as a Function of Average Work Trip Length for 18 Auto-Only Experiments

for example, will consume energy quite differently from one in which most business and industry is located along a beltway. Second, the transportation level of service must be taken into account. If the level of service is very low, for instance, traffic congestion, with its accompanying inefficient fuel use, will be worse. Finally, the role of public transit in the city under consideration must be determined. The percentage of trips made using transit may not be as important as whether transit is serving as a "safety valve" to relieve excess rush hour congestion.

Joseph Schofer and Robert Peskin have recently expanded and refined the work described above.[2] Based on their research for the U.S. Department of Transportation, they have concluded that poly-nucleated urban structures hold more promise for energy conservation than do other spatial arrangements.

The advantages of the polynucleated form are well illustrated by developments in some parts of the Minneapolis-St. Paul region, an

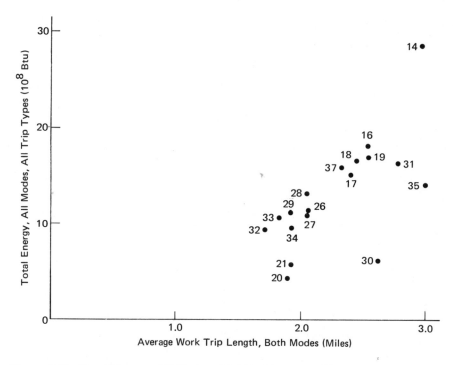

Figure 7-5. Total Energy, All Modes, All Trip Types as a Function of Average Work Trip Length over All Modes

area which seems to be moving toward the polynucleated pattern. There is a major regional shopping center in one subregion of the Twin Cities area, for example, that has a greater volume of annual retail sales than downtown St. Paul. It has all the characteristics of a mature central business district, providing many services that traditionally are available only in downtown areas. The presence of a wide range of service and employment opportunities in such a subregional center means that fewer trips downtown from the subregion should be necessary. Transportation energy consumption should therefore be less than it would be without the center. In a polynucleated city, perhaps proximity can be provided as a substitute for mobility.

Despite the extensive work already done, there are still many unanswered questions concerning the spatial structure-transportation energy relationship and the methodology used to examine it. Is the polynucleated city form really the most energy efficient? What actually happens to transportation energy consumption if population

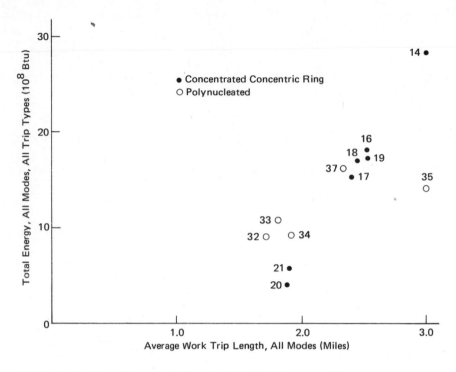

Figure 7-6. Total Energy, All Modes, All Trip Types as a Function of Average Work Trip Length, All Modes for Different Urban Forms

and employment are concentrated in outlying subregions? Is nonwork travel a better indicator of the transportation energy efficiency of alternative development patterns than work travel, the more traditional measure? Is the gravity model, used in previous research to predict choice of destination, really adequate for the task? Would a more disaggregated technique be better? Is is possible to model mode choice and trip generation rates as functions of land use, instead of stating those variables exogenously?

Research is currently under way to answer some of these questions, using the Twin Cities as a case study. For this study, the metropolitan area has been divided into a number of subregions (Figure 7-7) using the criteria listed in Table 7-1.

Table 7-2 describes "present" (1970) and "predicted" (1990) travel behavior for the subregions. It can be seen that in 1970 fully 60 percent of all trips other than work, and 36 percent of the work trips, were made to locations within the subregion in which they originated. It is predicted that these percentages will stay nearly the

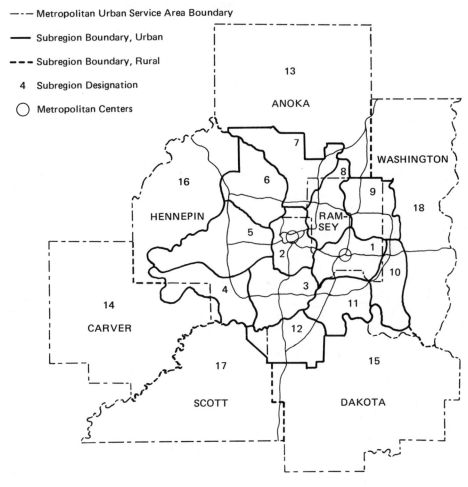

Figure 7-7. Transportation Planning Subregions, Twin Cities Metropolitan Area

same through 1990. The fact that so many nonwork trips presently are made within subregions suggests that people are quite willing to satisfy their nonwork travel needs close to home, thereby consuming less energy, if the necessary facilities are provided.

That nonwork trips constitute an important percentage of total travel is shown by data collected during the Travel Behavior Inventory conducted in 1970 in the Twin Cities area. Table 7-3 indicates that about half (46.6 percent) of total trips made in 1970 were for personal business, social and recreational activities, and shopping, and originated at home. There is a tremendous potential for energy

Table 7-1.   Criteria Used to Define Subregion Boundaries

1. Subregional boundaries should conform to the Metropolitan Urban Service Area defined in the Development Framework of the Twin Cities Metropolitan Council.
2. Subregional boundaries should conform to the extent possible with major natural features such as
   a. Mississippi River,
   b. Minnesota River,
   c. Minneapolis Park-Lake system,
   d. Lino Lakes-St. Paul water system,
   e. Lake Minnetonka.
3. Subregions should, to the extent possible, centrally encompass existing and proposed metroplitan highway links.
4. Each subregion should provide or be capable of providing major activities and opportunities, including
   a. A regional shopping center,
   b. Major employment concentrations,
   c. Significant residential development consistent with Development Framework population forecasts (between 100,000 and 200,000 persons),
   d. Sufficient health facilities and services,
   e. Social-recreational opportunities,
   f. Education opportunities.
5. Subregional boundaries should not separate residential neighborhoods or concentrations, commercial or industrial complexes, and close-in trade areas of major retail complexes.
6. Subregional boundaries should, where practical, conform with other administrative boundaries such as
   a. Political units,
   b. Municipal service districts,
   c. Data collection zones.

conservation if people make most of their daily nonwork trips within their respective subregions, rather than going beyond them.

A joint choice model is now being developed of trip frequency (0, 1, 2, or more than 2 trips per day), trip destination (subregional shopping center, elsewhere in the subregion, or outside of the subregion), and mode choice (bus, automobile with no passengers, or automobile with passengers), for all three of the nonwork trip types described above.[3] By using such a model, one can allow for the interaction of spatial arrangement and trip generation and it is no longer necessary to externally specify trip rates based on the respective locations of residences and shopping centers. The accessibility of a given destination from a particular residence is specified under this new scheme by a combination of the ratio of total travel time to trip distance and the ratio of the cost of the journey to annual income.

Once the simultaneous choice model is calibrated, the evaluation framework shown in Figure 7-8 will be followed. Alternative transportation systems and three different development scenarios will be specified for the subregions being considered. The computer will

Table 7-2. Present and Future Travel Behavior

A. Travel by Transportation Planning Subregions, 1970

Trip Orientation	Home-Based Work		All Other Trips		Total	
	Number	Percent	Number	Percent	Number	Percent
To CBD	189,863	17	267,128	6	456,991	9
Within Subregions	393,006	36	2,355,502	60	2,748,506	55
Between Subregions	515,865	47	1,294,146	34	1,810,811	36
Total Person-Trips	1,098,734	100	3,916,774	100	5,015,508	100

B. Travel by Transportation Planning Subregions, 1990

Trip Orientation	Home-Based Work		All Other Trips		Total	
	Number	Percent	Number	Percent	Number	Percent
To CBD	231,965	13	397,399	5	629,364	6
Within Subregions	623,054	35	4,009,809	55	4,632,863	52
Between Subregions	922,966	52	2,931,561	40	3,854,527	42
Total Person-Trips	1,777,985	100	7,338,769	100	9,116,754	100

Table 7-3. Travel Behavior Inventory, 1970: Internal Person-Trips by Residents[a]

Purpose	Number	Percent
HB Work	1,094,765	22.0
HB Personal Business	776,857	15.6
HB Social Recreation Other	808,952	16.2
HB Shop	735,158	14.8
HB School	277,351	5.6
HB Medical	63,626	1.3
HB Outdoor Recreation	187,948	3.8
N H B	1,032,068	20.7
Total	4,976,725	100.0

HB   = home-based trips
NHB = nonhome-based trips

[a]Total trips shown in this table differ from the totals in Table 7-2 because only resident-generated trips are shown here.

then determine the amount of energy consumed for nonwork travel under each of the alternative scenarios. Nonwork travel energy consumption will serve as a representative measure of the energy efficiency of each of the development schemes, and will make it possible to determine the most energy-conserving growth patterns for the Twin Cities area.

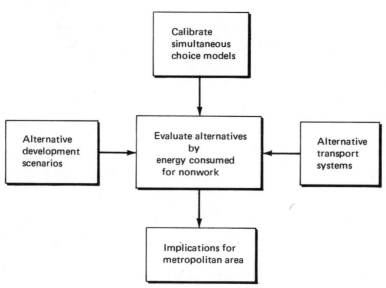

Figure 7-8 Evaluation Framework

NOTES

1. Lowry-type land use or spatial interaction models use basic or industrial and service employment information for a given urban area to allocate land uses (industrial, service, residential, transportation) to different zones within that area. They build upon the assumption that any addition to industrial employment will also cause an increase in the number of employees needed to provide services.

2. See Robert L. Peskin and Joseph L. Schofer, *The Impacts of Urban Transportation and Land Use Policies on Transportation Energy Consumption* (Springfield, Va.: National Technical Information Service, April, 1977).

3. This model is similar in concept to another recently developed by Thomas J. Adler and Moshe Ben-Akiva. See Thomas J. Adler and Mosha Ben-Akiva, "Joint-Choice Model for Frequency, Destination, and Travel Mode Shopping Trips," *Transportation Research Record*, no. 569 (1976): 136–150.

✲ *Part III*

Community Energy
Conservation in
Perspective

Chapter 8

Land Use and Energy Conservation: Is There a Linkage to Exploit?

Dale L. Keyes

Before discussing energy and land use, three points must be emphasized. First, energy, and more specifically the fuels used to produce energy, is one type of economic good. Fossil fuels or any other type of fuels have much in common with wheat, coffee, automobiles, or baseball bats. There is a supply, a demand, and a mechanism for matching the two. The mechanism may be a private market, a planned economy, or some mixture of the two. Whichever, it must be kept in mind that the energy crisis is a resource supply and demand problem. Energy should not take on a mystical quality.

Second, land use should be considered as one alternative strategy that may contribute to solving some of the problems related to energy. Land use is highly interrelated with other strategies; in fact, it is really a summary for more direct effects.

Third, it should be kept in mind that energy is but one land use consideration. Although a certain development pattern or a certain urban form may be the most energy efficient, it is not necessarily the best from other points of view. For example, it may lead to poor air quality. Energy is important, but we must keep it in perspective.

This chapter is based largely on research undertaken at the Urban Institute. It examines land use on a metropolitan-wide scale rather than discussing building orientation, site landscaping, or other small-scale features that hold promise for conserving energy. It also neglects the significant savings that can be achieved by the operation and maintenance function of local governments. What is discussed is the spatial configuration of land uses—the arrangement of urban activities in metropolitan areas. This is what urban planners have long called *urban form*, although that term may be somewhat misleading. In examining the shape of cities, we are also examining the arrangement and distribution of activities within those cities.

Conventional wisdom suggests that the most efficient city is a compact city, one in which a large proportion of the population lives in high-density residential areas and in which there is a great mixture of land uses. Development in such a city might well be constrained geographically. Efficiency should be increased when potential trip destinations are clustered together, for the length and number of trips will tend to be reduced. High densities should also increase the feasibility and efficiency of mass transit. Having mixtures of land uses theoretically should increase the amount of walking and the number of bicycle trips, both of which are even more efficient modes of travel than is mass transit. Space heating and cooling efficiency should also be greater in high-density dwellings, that is, in multi-family and townhouse units, than in single-family detached units. Higher densities generally mean fewer exterior walls, and fewer windows and doors through which heat can be lost.

Three questions can be asked at this point. First, what is the theoretical and empirical evidence—is the above view of an efficient city justified? Second, given that energy efficient urban structures can be identified, what is the magnitude of the differences in energy consumption between an efficient and an inefficient city? Third, what impact will designing new cities or accommodating new growth in efficient patterns have on the total amount of energy consumed in the United States over the next ten, twenty, thirty, or fifty years? Is changing land use patterns a realistic way to save energy? Even if certain urban forms are much more efficient than are others, can land use control or the manipulation of development patterns contribute significantly to reducing energy consumption?

Table 8-1 provides a useful starting point in answering these questions. The table shows the amount of energy consumed in the United States in 1968 by different economic sectors and by various end uses within each sector. This provides an indication of the potential for land use to influence total energy consumption.[1]

Residential space heating represents about 11 percent of total energy consumed and air conditioning about 0.7 percent; together they total about 12 percent. Out of that, about 8 or 9 percent is represented by residential development in metropolitan areas.[2] The transportation sector represents 25 percent of total consumption. About 14 percent of the total is the energy consumed by passenger travel, and about 8 or 9 percent of the total represents the energy consumed by urban passenger travel. Adding, it can be seen that

Table 8-1.   Total Fuel Energy Consumption in the United States in 1968 by End Use

End Use	Consumption[a] (trillions of Btu)	Percent of National Total
	Residential	
Space heating	6,675	11.0
Water heating	1,736	2.9
Cooking	637	1.1
Clothes drying	208	0.3
Refrigeration	369	1.1
Air conditioning	427	0.7
Other	1,241	2.1
Total	11,616	19.2
	Commercial	
Space heating	4,182	6.9
Water heating	653	1.1
Cooking	139	0.2
Refrigeration	670	1.1
Air conditioning	1,113	1.8
Foodstock	984	1.6
Other	1,025	1.7
Total	8,766	14.4
	Industrial	
Process steam	10,132	16.7
Electric drive	4,794	7.9
Electrolytic processes	705	1.2
Direct heat	6,929	11.5
Feedstock	2,202	3.6
Other	198	0.3
Total	24,960	41.2
	Transportation	
Fuel	15,038	24.9
Raw materials	146	0.3
Total	15,184	25.2
National Total	60,526	100.0

[a] Electric utility consumption allocated to each end use.

Source: Stanford Research Institute, *Patterns of Energy Consumption in the United States*, prepared for the Office of Science and Technology (Washington, D.C.: U.S. Government Printing Office, January, 1972): Tables 1, 6.

urban residential space heating and cooling, and urban passenger travel, together account for a very significant fraction, about 17 percent, of the energy consumed annually in the United States. These figures indicate that one is justified in examining land use as a means of conserving energy.

In the residential sector, a link has been identified between land use and energy consumed for space heating and cooling by residential structures; that link is building type. Because compact, high-density metropolitan areas typically contain multifamily structures and town-houses and very few single-family detached units, it makes sense to compare the amounts of energy consumed by each of these building types to see whether, by implication, more compact development is more efficient. There are a large number of approaches to this question, ranging from pure empirical analyses using data from individual struc-tures to simulation analyses based on engineering principles.

The results of recent research are illustrated in Figure 8–1. The horizontal axis indicates building density—from single-family detached units through small multifamily structures to skyscrapers. The verti-cal axis indicates therms consumed per square foot. (A therm is equal to 100,000 Btu.) By expressing energy consumption on a square foot basis, the influence of structure size on consumption is removed. The bottom curve on the figure, which goes all the way out to skyscraper, is based on data from all-electric homes in Chicago. Three different types of homes were studied, including single-family detached units, fairly large high-rise structures, and residential units in the John Hancock Building. These three examples span the range of structure types. The second curve from the bottom of the graph is based on an engineering simulation of prototypical buildings from the Washing-ton, D.C. metropolitan area. In such a simulation study, the researchers specify such factors as the size of the building, the con-struction, the thermal insulation characteristics, the building orienta-tion, and the climate. They then estimate the thermal loads. In this study, the researchers examined a single-family detached dwelling, a townhouse, a small garden apartment, and a small high-rise. The third curve from the bottom is based on statistical work by the Urban In-stitute using survey data provided by the Center for Metropolitan Studies in Washington, D.C. The fourth and highest curve is based on additional work with data from the New York City region. In this case, the unit of analysis was the county rather than the household, but the same approach was used.

The first aspect of the graph that should be noted is that there is a fairly wide scatter of points for any building type. For example, de-pending on the source of the data, a single-family detached unit uses anywhere from 0.7 to about 1.5 therms per square foot. This is not surprising considering the various methodological approaches that were used, and, more importantly, the different parts of the country from which the data were taken.

The next aspect to be noted is the consistency in the general slope of the curves. The curves indicate the relative change in energy use with a change from low-density to higher density structures. The

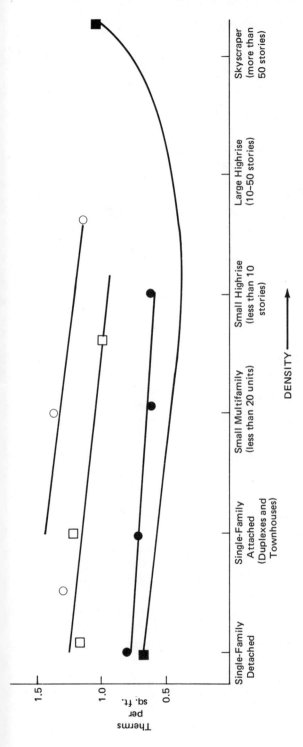

○ Regional Plan Association, Inc. and Resources for the Future, *Regional Energy Consumption* (RPA Bulletin 121, New York: Regional Plan Association, January 1974).

□ Response Analysis Corporation, *Lifestyles and Energy 1973 National Surveys*, Report Prepared for the Washington Center for Metropolitan Studies (Washington, D.C.: Washington Center for Metropolitan Studies, December 1974).

● R. W. Anderson, *Residential Energy Consumption, Single Family Housing* (Columbia, Md.: Hittman Associates, Inc., March 1973) and M. Tokmanhekin and D. G. Harvey, *Residential Energy Consumption, Multi-family Housing Final Report* (Columbia, Md.: Hittman Associates, Inc., June 1974).

■ A. L. Sweet, "Effects of Residential Building Type on Energy Consumption," *Building Research* (April/June 1974): 18-24.

Source: Dale L. Keyes and George E. Peterson, "Metropolitan Development and Energy Consumption, Land Use Center Working Paper 5049-15" (Washington, D.C.: The Urban Institute, March 1977).

Figure 8-1. Relative Energy Efficiency by Type of Dwelling

slopes consistently indicate that units in small high-rise buildings are 30 to 40 percent more efficient than are single-family detached homes.

A third point is illustrated by the bottom curve on the figure. Energy consumption per square foot in the John Hancock Building (indicated by the point at the far right end of the curve) is higher than the consumption of the single-family detached unit (at the far left end of the same curve). These data suggest that beyond some threshold number of floors in a multifamily structure, the amount of energy that is needed to run the elevators, to light and heat the common areas, and to transport water and waste material vertically outweighs the thermal efficiency advantages of the individual units. There is some threshold, then, beyond which increasing density (i.e., increasing the size of multifamily structures) leads to energy inefficiencies. The more important point, however, is that, as noted above, there is a relative savings in energy consumption of about 30 to 40 percent as one moves from single-family to higher densities.

A similar approach was taken in analyzing transportation energy consumption. A number of simulation and empirical studies relating travel patterns to urban development patterns were examined. In addition, gasoline consumption data were related to various characteristics of metropolitan areas. Generally, the conventional wisdom was well supported. Compact, high-density cities tend to be more efficient; they require less travel. However, it was also found that efficient use of transportation energy is also possible with other spatial arrangements. For example, if a city sprawls both in its population distribution and in its employment distribution, there does not appear to be a great increase in travel.

The next question to be examined is the magnitude of differences between efficient and inefficient cities in terms of both residential and transportation energy use. Rather than examining the extremes of urban form postulated in a number of simulation studies, it is more useful to ask, "What is realistic?" The single-family detached home cannot be eliminated and the automobile cannot be outlawed in the name of energy conservation.

In order to predict changes in residential consumption, we started with the distribution by building types of new units constructed during the period 1973 to 1975. Before that period, many single-family detached homes but very few high-rise structures were built. We assumed that this distribution would change, that there would be a rather significant movement away from single-family detached homes to townhouses and high-rise dwellings. This assumed distribution may in fact, be overly optimistic; people may not abandon the single-family home to the extent postulated. But the assumptions are much more realistic than most others that appear in the literature. Energy efficiency factors for different dwelling unit types were applied to the assumed distribution. It was found that the difference between the amount of energy that would be consumed by new growth if present trends continue, and the amount that would be consumed with this assumed new distribution was about 10 percent. That is, the energy consumption added by new people moving into an area would be 10 percent higher if they lived in buildings characteristic of those built in the early 1970s, than if those people favored more efficient homes to a realistic degree.

A similar approach was used to determine travel savings. Various simulation analyses were examined and the most realistic scenarios were identified. In addition, statistical work involving gasoline sales data was used, together with what were believed to be reasonable changes in the land use variables. These analyses indicated that savings of between 10 and 15 percent in the rate of increase of transportation energy consumption was possible by more efficient accommodation of new growth.

The final question posed is this: What do these findings mean in terms of total energy use in some future year? The year selected was 1985, a convenient year because the Federal Energy Administration (FEA) developed their Project Independence forecasts for 1985 as well. The analysis indicated that the savings in annual energy consumption made possible by accommodating new growth more efficiently would be around 0.35 percent of the national total. This is obviously not an overwhelmingly large saving.

This reduction in energy use was compared to those possible by two other strategies used by the Federal Energy Administration in making their assessments. As is shown in Table 8-2, the FEA predicted that an increase in gasoline prices of 29 cents per gallon from 1972 to 1985 (an 80 percent increase) would yield savings, expressed as a percentage of the total consumption in 1985, of about 0.4 percent. Similarly, it was predicted that a mandatory automobile fuel efficiency standard of 20 miles per gallon would bring about savings of approximately 2 percent. Although these percentages are small, it should be kept in mind that in absolute quantities they represent quadrillions of Btu. A small percentage is still a sizable saving in an absolute sense.

These comparisons do not make one very sanguine about the potential of land use controls or the manipulation of development patterns to make a significant impact on energy conservation. It is much more reasonable to consider the regulation of land uses as a

Table 8–2.   Effectiveness of Energy Conservation Strategies

Strategy	Percent Savings in Total U.S. 1985 Energy Consumption
Modified growth pattern	0.35
Increased gasoline prices (from $0.36 to $0.65)	0.40
Mandatory 20 mpg efficiency standard	1.90

facilitating mechanism. If the demand for higher density, more compact development patterns increases because of higher energy prices or for other reasons, local governments should then design their zoning ordinances to accommodate this increased demand. If in fact there is too little land zoned for higher density development, then it is encumbent upon local governments to make more land available. Likewise, local governments should look very carefully at the money-lending market. If inner-city areas are "red-lined" so that centrally located sites cannot receive construction money, then an attempt should be made to remove those constraints. Facilitating energy conservation through land use decisions should be encouraged, so long as it is remembered that land use changes are not the main solution to energy problems. Land use manipulations alone would appear to be an indirect, rather cumbersome, and not terribly effective means toward an energy conservation end.

NOTES

1. Because the information in Table 8–1 is somewhat dated, the actual numbers should not be taken literally, but the relative differences are approximately correct for today.

2. It should be noted at this point that commercial space heating represents about 6.9 percent of the total, and commercial air conditioning about 1.8 percent. Although this is useful information and is very relevant to land use issues, it will not be discussed further. Data are very scarce on buildings in the commercial sector, and the data that are available do not reveal much of a trend in energy consumption by different types of commercial buildings. Therefore, the remainder of this chapter focuses on residential and transportation energy consumption, where data are more numerous and reliable.

✳ *Chapter 9*

The Law and Energy Conservation

Grant P. Thompson

This chapter focuses on three aspects of the law in relation to energy conservation. First, it discusses the limitations of law in a technical area. Second, it considers the ability of law to affect neighborhood design to conserve energy. Third, it examines the use of law to encourage energy conservation in individual building design. The ideas and information presented grew out of a three-year study, funded by the National Science Foundation, of "State and Local Energy Conservation," conducted through the Environmental Law Institute. Of course the opinions expressed are solely those of the author, not NSF or ELI.

LAW IN A TECHNICAL AREA

Many people have a rather awestruck idea of what lawyers do and of the influence lawyers have in our society. Indeed, lawyers who work in private practice and in public occupations do have tremendous power; they organize and structure many of the ways we deal with each other. Nonetheless, we should never forget that law is not an independent source of knowledge and information about what a society ought to do. Instead, law is simply a tool; it is the bone, the structure, the set of rules that a society agrees upon. An individual law has virtually no content that arises from the discipline itself. Instead, the content of laws is drawn from engineering, science, social sciences, economics, and other fields. Laws simply try to carry out ideas developed in these other fields.

On the other hand, it would be disingenuous to suggest that the discipline of law adds nothing to the subjects that it regulates. Law has a subtle molding effect as it tends to color the way we frame questions in our society. Anglo-Saxon law, for example, views the adversary process as the natural one for solving problems. Under this influence, we frame government regulations or rules in our society to emphasize "A versus B," only rarely considering the possibility that A and B might work together or be similarly motivated in some way to solve their joint problem. Law worries about administrative ease in solving a problem. No matter how good a scheme is, if it is too complicated for human beings to understand, they will not follow it. Law worries about equity. If what is proposed is not fair, in the long run people will not do it. It worries about established categories of property and about traditional rights that people feel they have. Law is strongly in favor of stability as a social goal; it suggests that society move slowly and that there is virtue in doing so. These characteristics of law, plus the fact that law is only a tool of society, have a tremendous effect on what can be done by means of law to change the way energy is used.

After spending three years trying to devise clever ways to get people to do things that would be good for them, one feels a considerable sense of frustration about the use of law to enforce good works. The mandatory controls over energy use proposed in many quarters ignore what may be the most important tool for energy conservation—the force of self-interest and greed. If a homeowner is told that there are certain things that he must do to his house—such as insulate, install storm windows, or caulk—the homeowner will be willing to comply with these requests so long as he will save money by doing so. But he may not do enough insulating, storm windowing, or caulking from the nation's point of view because the price of energy in this country generally does not reflect the cost of finding new supplies. From the homeowner's point of view R–30 insulation might be sufficient, but the nation, looking at the replacement cost of energy, realizes that an R–40 standard is required. If the government tries to use law to regulate the homeowner, telling him in effect that R–40 is good for all of us, the same unsatisfactory result that has occured with the 55-mile-an-hour speed limit may well be achieved. People tend to evade the law when it costs less to do so than to obey it. Thus, our work has led us to believe that law alone cannot bear the burden of energy conservation in this country. The way to get this country moving toward energy conservation goals is to raise the price of energy to its true replacement cost, including environmental and social costs. Having said that, however, it is clear that such a course will not be followed in the next few years, and may never be followed in an unalloyed fashion.

LAW FOR AFFECTING
NEIGHBORHOOD DESIGN

How can the law affect the way that neighborhoods and small city units are developed? Very little is known about the exact relationship of energy use to urban design. We do know that there are two goals that seem worth pursuing. First, because urban design affects the transportation energy that people use, one ought to strive for compact, contiguous urban design, with jobs and small shopping centers close to homes. Second, because energy can be saved by increasing the number of party walls, more multifamily housing should be built. In spite of the apparent clarity of these two goals, we do not really have a very clear idea of the best way to shape our cities to save energy. To make matters more complex, in our society housing meets psychic as well as physical needs. "Housing for every American" is the rallying cry, for decent housing represents the hopes and dreams of most citizens of moving up economically in the society. That psychic goal may bump squarely into national efforts that concentrate solely on energy-saving designs for buildings.

Given the constraints of uncertainty, of a lack of data, and of a lack of any true conviction as to what the nation's social goals should be in housing and urban design, our research tried to suggest laws and policies that would gently lead people to conserve energy. Perhaps not surprisingly, it turned out that there are not very many new ideas. The same legal controls that have been talked about for other social purposes can also be used to encourage compact and contiguous multifamily dwellings. Examples of such traditional land use planning devices are site planning reviews that award added points for energy conserving urban form, and density bonus schemes that permit developers who propose energy-conserving site and building plans to have more units per acre than the ordinary zoning allows. Denser housing can be encouraged closer to the center of a city by announcing long in advance that city services will not be extended beyond a certain point. Thus, even though land use regulations are being used for a different purpose, the legal strategies available are really the same familiar menu of planning tools. A discussion of these devices, together with actual examples of land use planning being used for energy-conserving purposes, are contained in *Using Land to Save Energy*.[1]

Planning for urban form could benefit from further studies that explain exactly the relationship between form and energy use. Research is needed to determine what an energy-conserving community really would look like. We need to study how energy is used, what type of community will attract people to work in that community rather than commute to distant jobs, and what kinds of social, economic, and legal incentives can be built into that community to encourage or force residents into energy-conserving personal habits. Armed with this research information, the law could act more rationally in helping shape an energy-conserving city.

LAW AND BUILDING DESIGN

The chapters by David Harrje and Michael Sizemore demonstrate that many good ideas have been suggested for saving energy in buildings. Why are they not being put into practice? How can people be made to use them? One reason they are not being used is because they cost more than they pay back to the building owner. A second reason is that many buildings are built by one person to be sold or rented as soon as possible to another person, and it is the second person who pays for the energy used to operate the building. This situation drives builders and developers to minimize the initial cost of structures, sacrificing operating efficiency. Although some of the energy-conserving devices that have been suggested have short payback periods (of less than two or three years) they are frequently somewhat more expensive to purchase, and therefore do not appeal to builders who want to keep the initial price low in order to sell or rent quickly. A third factor that limits the implementation of energy conservation ideas is simply a lack of information. Until recently, not very many people were interested in saving energy, for Americans apparently believed that it would last forever.

Tradition and inertia may be the most powerful forces working against energy conservation. Builders argue that there are benefits to building houses the way that they and the unions know and understand. If they have to build homes in some other way, they will encounter uncertainties that will add to the cost. Much of this tradition is "frozen" into law and regulation. Building codes, minimum property standards put out by the federal government, handbooks for smaller builders, and certification programs of the appliance manufacturers' associations all carry the force of law. There is a tremendous resistance to change in the legal system.

Clearly what is needed is some means of making energy conservation in building construction part of the usual course of doing business. The most obvious existing legal mechanism for mandating conservation is to rely on the traditional building code mechanism, but to add an energy conservation component. In principle, there are two ways that one can approach an energy conservation standard. One method can be called the budget approach, and the second can be called the cookbook approach. A budget approach simply mandates that a given building may not use more than a stated amount of energy in its operations; the cookbook approach mandates the use of certain approaches or technical fixes to achieve energy savings. There need be no difference in the energy goal no matter which approach is used, but there may be some important differences in practicalities, and it is to those differences we shall now turn.

Under the budget approach, one is not concerned with how the building designer goes about saving energy. The only concern is to make certain that new buildings do not exceed a certain amount of energy consumption per square foot. Various groups have proposed alternative energy budget approaches. There is the so-called pure budget approach, under which an adequate energy budget for a building is specified and then adjusted for climate and usage. There is also a modified form, which takes this pure idea and simplifies it enormously. One might choose some standard—such as the one promulgated by the American Society of Heating, Refrigeration, and Air Conditioning Engineers in 1975 (ASHRAE Standard 90–75)—and state that a building must use no more energy than a building would that was built according to that standard. The federal government, under the requirements of law, is now developing an energy budget approach that will become mandatory for the states to adopt. Unfortunately, comparatively little information was available concerning the proposal in mid-1978.

The major technical difficulty encountered for anyone proposing an energy budget approach is to set realistic energy consumption figures for a given building type and climate. This country has almost no experience in determining what constitutes a reasonable amount of energy to use for different applications. The American Institute of Architects has suggested one solution to this dilemma. The AIA recommended that we determine the energy consumed by buildings built in a particular area over the previous three years, find the median figures, and make that the budget for the next year. Each year the figures would be adjusted downward in order to remain current with changes in actual architectural practice.

A second technical problem with the budget approach is that the building designer has very little actual control over how a building will be operated. He does not know in advance whether an office building will contain dozens of copying machines, which pour out heat all day long, or whether it will have a computer, or whether people will be working on weekends. These uncertainties make it difficult to develop an accurate budget.

From a legal point of view, there are two problems with the budget approach. First, there is the question of ease of administration. The energy budget approach, particularly in its pure form, tends to require complicated calculations. Indeed, the computations needed are far more complex than those that form the usual diet of an average building inspector. Thus, before a system can be established, either careful training will be required or some centralized system will have to be developed to process applications. Second, questions of legal guarantees may arise when the energy consumption level of buildings is stated in advance and is part of the permitting process. Suppose that the prospective owner asks for a guarantee that the new building will conform with the standards? Such a guarantee is not unreasonable, but will a builder give it? What if the builder does not give such a guarantee and the building fails to meet the standard. Who pays? The owner has been relying on the building department, the building department has been relying on what the architects said, the architects have been relying on the contractor, and so on. The building is unlawful, but it cannot be torn down. Who is responsible for taking care of the problem? Does the new owner have cause for suit for excess operating costs? All of these problems can be solved, but they are a deterrent to the approach that puts a specific performance standard into law.

The major advantage of the budget approach is its flexibility. It permits the designer to spend energy lavishly in one part of a building, be parsimonious in another, and yet still meet the budget overall. However, more technical development of the budget method is needed. Realistic budgets must be developed. More administrative work must be done so that this technique can be readily administered by people without technical training. One may hope that the new federal standards will take these concerns into account.

The second general method of enforcing energy conservation by law is the cookbook approach. Using this approach, one collects a series of good energy-saving ideas, such as caulking windows and insulating walls and ceilings to a certain R value, and then enacts this list into law. No permit may be granted unless the builder can show the building inspector that he has complied with all of the points on the list.

There are problems with this method as well. The detailed lists may begin to look inappropriate soon after they are adopted. Items on the list may be appropriate for another climate, or perhaps a builder has done something different in the design of his structure or mechanical systems that makes a certain requirement a bad idea. It is very difficult to deal with items on an inappropriate list that has the force of law.

One approach is to give the building inspection department discretion to make changes in the list. Unfortunately, American administrative law tends to be very wary of giving administrators much discretion. When administrative agencies are given discretion, they are also subject to suits from persons or companies who think that the discretion is being abused. Such suits slow the process and decrease easy administration.

The great advantage of the cookbook approach is that it is simple to understand and simple to administer. It is easy to see whether or not a structure has a certain amount of insulation: you simply go out and read the label. This is in contrast to trying to figure out in advance whether a building of a certain design will meet a budget standard of a certain number of Btu used per square foot on a year-long average. Perhaps for these reasons, the early development of energy conservation standards in this country has favored the cookbook approach. This seems to have been a wise approach, embracing the simpler methodology while researchers are learning more about the systems analysis or budget approach. The predominately cookbook approach of ASHRAE Standard 90-75 has been widely adopted (under the pressure of a federal energy conservation law), and the newer budget approach will then supplant it.

CONCLUSION

Our society is still in the very early stages of deciding what regulations we will impose on ourselves for building energy conservation. We are still nervous about the enterprise, because we do not like to tell people what they can do with their own homes and money. There is a healthy American tradition of distrust of government and dislike of picayune regulations. However, as energy prices continue their rise in the future, it is inevitable that we will see much more regulation and much more technical tightening of standards. It will be an exciting time for lawyers, architects, and engineers working together in an attempt to deal with these problems of energy use and translating those technical findings into sound, efficient, and clear legal directives.

NOTE

1. Corbin Crews Harwood, *Using Land to Save Energy* (Cambridge, Mass.: Ballinger, 1977).

Chapter 10

Conserving Energy
Efficiently

Bruce Hannon

Why is energy conservation important? Figure 10-1 compares U.S. energy and income distribution in 1960-1961 (the latest year for which figures on income are available from the Bureau of Labor Statistics) to a uniform distribution and to the world distribution of energy consumption. As time has passed, U.S. income distribution has become increasingly equal; the 1960-1961 curve is much closer to the uniform distribution than was the curve during the Great Depression, for example. In 1960-1961 energy was distributed over the population somewhat more evenly than was income, perhaps indicating that energy was underpriced.[1]

In the case of world energy distribution, however, it is evident that a situation of great inequality exists. China and India, which represent 50 percent of the world population, consume only 5 percent of the energy. In contrast, the United States, with 5 to 7 percent of the population, consumes 30 to 40 percent of the energy. As the rest of the world awakens to economic development and to this magic elixir called energy, and as other countries realize that energy is the basis for improving national life, they will demand increasing quantities of fuels. They will consume energy at a much faster rate than we have ever conceived of doing. This will drive up prices in the United States, and will cause curtailments and international conflict. Energy must be conserved if there is to be any hope of avoiding these consequences. In addition, energy must be conserved to protect the environment. There is a direct correlation between fossil fuel energy use and environmental degradation.

Having established that there is a need for energy conservation, current energy consumption patterns must be closely examined to

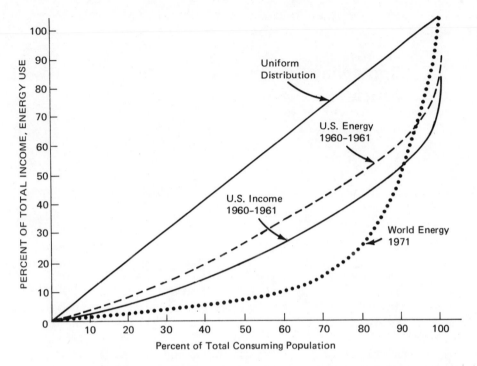

Source: Bruce Hammon, "Energy, Land and Equity," in Wildlife Management Institute, *Forty-first North American Wildlife Conference* (Washington, D.C., Wildlife Management Institute, March 1976): 271.

Figure 10-1. Distribution of U.S. Consumer Income after Taxes, Energy Use, and World Energy Use

determine how that conservation can best be accomplished. As a first step, the flow of energy through the entire United States economy on a detailed level was modeled, using an input-output model of the economy constructed by the United States Department of Commerce.[2] The Department of Commerce used industrial and commercial data collected by the Bureau of the Census to build its model, which expresses economic activity as flows of dollars between 400 different economic sectors. The results of that modeling were converted into energy flows, as though energy were the currency of the economy. The quantities of energy of various types consumed were then distributed over all the purchased goods in the economy. Goods purchased for capital formation, for direct consumption, for export, and for government use were all included.

Table 10-1 illustrates the allocation of energy, by type and by category or activity of consumption, that was developed. The figures for Total Primary Energy consumption indicate that about two-thirds of the energy used directly and indirectly goes for personal consumption. Roughly half of the energy consumed in this way is consumed directly, whereas the rest is embodied in purchased goods and services.

The same sorts of calculations were made for employment, in order to be able to compare the employment distribution to the energy consumption distribution. Estimates of the person-hours worked in the country were passed through the model in order to see how they are directly and indirectly distributed. It was found that about 60 percent of all employment is absorbed by the activities of personal consumption, compared to 63 percent of the total energy. As another example, federal defense accounts for 8 to 8.6 percent of the total labor consumption and 6 to 6.6 percent of the energy consumption.

Figure 10-2 shows the energy and employment intensities of each of 400 sectors of the economy. The chart illustrates the rates at which various industries, as they produce dollars of output, demand from the economy energy and labor, both directly and indirectly. As the chart indicates, certain industries, such as asphalt and aluminum manufacturers, need very large amounts of energy, but use very little labor, whereas others, such as credit agencies, tobacco, hotels, and hospitals, are very labor intensive, but use little energy. It is important to note that, according to the data, industries can be either labor-intensive, or energy-intensive, but not both. A chart of this sort was also constructed with 1967 data. This chart indicates that industries had become slightly more energy intensive and a little less labor intensive than they were in 1963.

Once the present situation in regard to energy consumption is known, one can next ask, What will happen as the price of energy is raised? In order to answer this question, it can be assumed that, in a very general sense, labor, capital, and energy are the ingredients needed to make any product. (For present purposes, the land itself, usually considered as another factor of production, is ignored.) The prices of these commodities dictate the ways in which they will be used in the economy. The ratios of the prices of labor to energy, of labor to capital, and of capital to energy determine whether and how fast one might switch from a more energy intensive and less labor intensive economy, for example, to one that is more labor intensive and less energy intensive.

Figure 10-3 illustrates the changes in the ratios of the costs of capital (the interest rate of AAA industrial bonds), of energy (the price of electricity), and of labor (the salary of the average industrial worker) that occurred from 1925 to 1975. The wage-electricity ratio

Table 10-1. Percentages of Total[a] Energy and Employment (Direct and Embodied) for Final Consumption Activities, 1963 and 1967

Activity	Coal		Refined Petroleum		Electricity		Gas		Total Primary Energy[b]		Total Employment	
	1963	1967	1963	1967	1963	1967	1963	1967	1963	1967	1963	1967
Personal Consumption	52.0	49.9	71.3	69.2	67.4	65.7	70.5	65.4	66.2	63.0	62.5	59.5
Capital Formation	15.7	14.4	7.3	7.2	11.3	10.7	11.5	10.8	10.7	10.1	11.8	12.3
Inventory Change	1.3	5.2	1.5	2.0	1.0	1.7	1.0	2.8	1.3	3.1	1.0	1.4
Exports	16.1	14.0	6.7	6.1	5.4	5.4	5.2	6.0	8.6	8.3	4.0	4.1
Federal Defense	6.6	6.8	6.2	7.7	6.6	5.8	4.8	5.1	6.0	6.6	8.0	8.6
Federal Other	1.1	1.6	1.2	1.5	0.6	1.4	1.1	1.6	1.2	1.6	2.6	2.5
State and Local Governments	7.2	7.9	5.7	6.3	7.5	9.3	5.8	8.3	6.1	7.5	10.1	11.6
Total Final Demand[d]	100.0	100.0	100.0	100.0	100.0	100.0	100.0	100.0	100.0	100.0	100.0	100.0

[a] Domestic base: All figures contain estimates of energy and labor embodied in imports.

[b] Coal plus crude petroleum plus natural gas plus the hydro power portion of electricity converted at the fossil conversion rate.

[c] Jobs basis; includes household and government workers; excludes military personnel.

[d] Some columns may not add to 100 percent because of rounding.

Source: Bruce Hannon, "Energy, Growth and Altruism" (Urbana: Center for Advanced Computation, University of Illinois at Urbana-Champaign, October 21, 1975): 30.

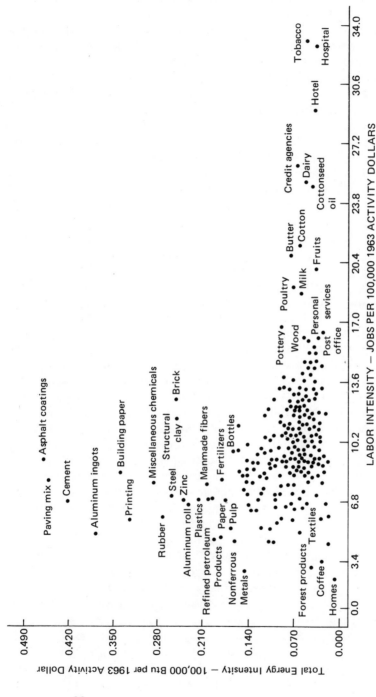

Source: Bruce Hannon, "Energy, Growth and Altruism "(Urbana: Center for Advanced Computation University of Illinois at Urbana-Champaign, October 21, 1975): 30.

Figure 10-2. Total (Direct and Indirect) Energy and Employment Intensities for All U.S. Industrial Sectors in 1963

83

increased from 1926 to 1972, implying that labor increased in cost relative to energy. Similarly, until about 1950, the wage rate rose faster than the cost of capital. The price of labor was rising so fast that it was literally driving itself out of the market. There was a continual, increasing pressure to substitute energy and capital for labor by the classical scheme of automation.

In 1950, the Federal Reserve Board gained control of the bond interest rate. The board's actions have since caused a standoff between the rate at which wages have risen and the rate at which the price of capital has gone up, stabilizing the wage-capital ratio.

In about 1950, the ratio of the cost of capital to the cost of electricity also began to increase, creating a tendency to use less capital and more energy. The behavior of the electric utility industry illustrates this change. As capital became more expensive than electricity in the early 1950s, utility companies began extensive campaigns to encourage the use of air conditioning. At that time, it was less expensive for the companies to generate electricity than to allow generating equipment (capital) to remain partially idle during the summer. As use of air conditioning increased, the peak electricity loads began to occur in the summer instead of the winter. Because the utility companies were still faced with a high capital cost, they began a campaign for increased installation of electric heating, so that they could once more achieve balanced, complete utilization of their generating equipment. With the Arab oil embargo, there was an abrupt end to the upward trend in the wage cost-electricity cost and capital cost-electricity cost ratios. Since that time, electricity has become increasingly expensive relative to labor and capital, and the utility companies have begun campaigns to encourage energy conservation.

These trends in electricity prices hold true for other fuels as well. As energy becomes increasingly expensive relative to the other factors of production, there will be a growing tendency to substitute additional labor and capital for energy in production processes. Labor will probably also be substituted for capital, for capital takes energy to make.

Now that it has been established that there is a trend toward less energy-intensive production, who will resist this trend can be predicted. Unions seem especially likely to oppose such a shift, particularly high-wage unions whose members work with very automated processes in large centralized factories. The trends toward the use of larger quantities of labor at lower wages and toward less centralized manufacturing (to save on transportation energy) can only hurt their members. Investors will also resist the trend toward a less efficient, less centralized use of capital, for less intensive capital use means a lower return on investments.

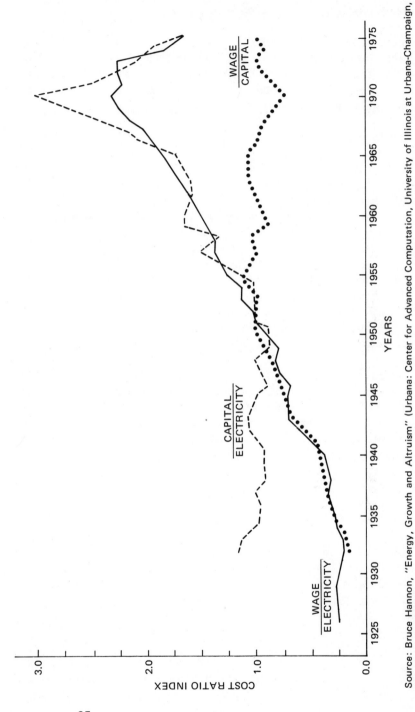

Source: Bruce Hannon, "Energy, Growth and Altruism" (Urbana: Center for Advanced Computation, University of Illinois at Urbana-Champaign, October 21, 1975): 39.

Figure 10-3. U.S. Industrial Labor, Electricity, and Capital Cost Ratios, 1926 to 1975, Current Dollars, (Manufacturing Worker's Hourly Wage, the Industrial Price of a Kilowatt-Hour of Electricity and the Yields on AAA Corporate Bonds).

A consumer might try to circumvent the fact that energy prices are rising faster than current, or paycheck, wages by working more hours. Unfortunately, as will be shown, this will not solve the dilemma. There is a high correlation between increases in income and increases in the amount of energy consumed.

Figure 10-4 illustrates, using 1961 data (the latest year for which information is available), that direct energy purchases (the bottom curve on the figure) including gasoline, home heating fuel, electricity, and so on, tend to saturate with a sufficient increase in income. Beyond a certain point, one just cannot drive any more cars and cannot heat and keep the lights on in any more rooms. This curve illustrates the idea that any increase in the tax on direct fuels will disproportionately hurt the poor. Because of the saturation effect, the rich spend a much smaller proportion of their incomes on direct energy than do the poor. As much as two-thirds of poor people's energy purchases are direct, whereas as little as one-third of the energy bought by the rich is in this category.

The close relationship between energy consumption and income does not end, however, once income rises above that level at which direct energy use is saturated. There is energy embodied in the non-energy goods, such as homes, cars, food, furniture, movies, and bus and airplane tickets, purchased by higher income groups. As Figure 10-4 illustrates, the wealthy consume enough energy indirectly to allow total energy consumption to increase almost linearly with income. Thus, if a person works longer or increases his investment and thereby doubles his income, he will double his consumption of increasingly expensive energy as well. City dwellers who assert that they use less energy because they live in apartments and ride the subway are incorrect. If their incomes are 25 percent higher than the average, their energy consumption is also 25 percent higher than the average. Their greater incomes allow them to consume more energy indirectly.

Other calculations using this consumption data indicate that the upper middle income class in the United States, which includes people with incomes about 40 percent higher than average, consumes more energy than any other income group. The upper middle class life-style is such that it results in a very high individual energy consumption level. This characteristic, combined with the very large number of people in this category, explains the high total energy consumption by this class. These high individual and class consumption levels seem even more important when one realizes that the upper middle class is the group most average income people aspire to join.

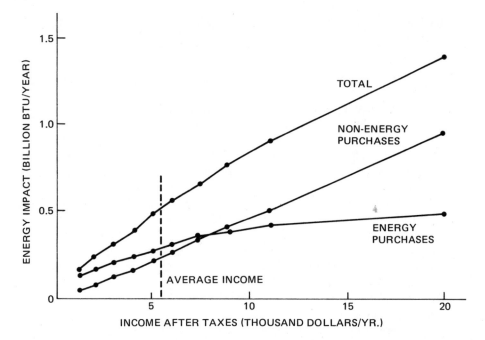

Source: Bruce Hannon, "Energy, Land and Equity," in Wildlife Management Institute, *Forty-first North American Wildlife Conference* (Washington, D.C., Wildlife Management Institute, March 1976): 271.

Figure 10-4. Direct, Indirect, and Total Consumer Energy Impact Plotted Against Consumer Income, 1960 and 1961.

This analysis of energy consumption suggests that energy be taxed at its source. Coal should be taxed at the mine, petroleum at the wellhead, and hydroelectric power at the dam. The amount of the tax should be based on the amount of energy being brought into the system. With such a tax, the increased energy costs would be spread throughout the economy and would affect all items of personal consumption according to the amount of direct or embodied energy contained in those items. Because there is a nearly linear connection between income and the total amount of direct and indirect energy consumed, a source tax on energy would be relatively equitable. Such a tax would also prevent private energy companies from collecting windfall profits from energy price increases and investing those profits outside the United States.

The amount of money raised by a source, or severance, tax could be as large as $30 billion per year, assuming that the tax was sufficient to increase the price of energy by about $1.30 per million Btu.

Part of this considerable sum might be used to pacify the companies that objected to the tax. The rest could be redistributed as a tax rebate, perhaps with a special emphasis on those people who have few, if any, options.

If it is impossible to avoid the dilemma of rapidly rising energy prices by earning more money, one might try instead to practice energy-conserving personal habits such as living in a smaller house or riding a bicycle instead of driving a car. Unfortunately, this strategy is unlikely to result in a true reduction in energy consumption. Any savings in energy costs achieved by reductions in direct consumption will probably be negated by the purchase of other goods with large amounts of embodied energy.

The difficulty of using less energy can be illustrated with a transportation example. Table 10-2 contains information that has been developed about the monetary and direct and indirect energy costs of intercity and urban travel by various transportation modes. Also included is a calculation of how many jobs are needed for each one million passenger miles of travel by each mode.[3] The table indicates that the electric commuter train is the most energy intensive mode of intercity travel, consuming 9,900 Btu per passenger mile. This high consumption level is the result of its poor load factor and of the fact that it uses electricity. Airplanes consume nearly as much energy (9,800 Btu) as commuter trains, whereas automobiles on the average use about 5,900 Btu per passenger mile. Regular trains and buses consume only 4,000 and 2,700 Btu, respectively. Both cars and buses consume more energy in urban than in intercity driving, because of stop-and-go traffic. Urban motorcycles also use a large amount of energy. The energy required by bicycling and walking is supplied by the agricultural fuel and fertilizer used to produce the extra food consumed by people engaging in these activities.

If one were to change from bus riding to bicycling, the cost of going a thousand miles would be reduced by $79, according to the chart. Table 10-3 indicates that such a change would result in a saving of 51,000 Btu for each dollar saved. Assume that the money saved from such a switch could be spent on any of the items shown in Table 10-4. This table lists the amount of energy embodied in each dollar's worth of a number of items commonly purchased by Americans. If the savings were spent on men's clothing, for example, there would be a net reduction in energy consumption, for the amount of energy embodied in one dollar's worth of clothing (31,400 Btu) is less than the amount of energy saved by going from bus to bicycle (51,000 Btu). On the other hand, if the money saved was spent on kitchen appliances, which represent 58,700 Btu per

Table 10-2. Selected Results on the Total Dollar, Energy, and Labor Impacts of Consumer Options in Transportation During 1971. Data are Expressed in Terms of the Requirements to Move 1 Million Passengers 1 Mile

Transportation	Load Factor	Thousands of Dollars (1971)	Energy[a] Millions Btu	Direct[a] (%)	Jobs
		Intercity Transportation			
Car	2.9 people	55	5,900	51	3.7
Plane	53% full	58	9,800	73	3.8
Bus	47% full	39	2,700	51	3.1
Train	37% full	44	4,000	58	7.2
Electric commuter[b]	31% full	128	9,900	11	8.5
		Urban Transportation			
Car	1.9 people	69	8,900	58	4.2
Bus	12.0 people	105	5,300	57	8.3
Motorcycle	1.1 people	57	4,200	49	1.6
Bicycle	1.0 people	26	1,300	59	1.7
Walking	1.0 people	NA	710	51	0.7

[a] Vehicle transportation fuel only.
[b] The PATH Commuter System, New York-New Jersey, 1971.
Source: Bruce Hannon, "Energy, Growth and Altruism" (Urbana: Center for Advanced Computation, University of Illinois at Urbana-Champaign, October 21, 1975): 39.

Table 10-3. Urban Transportation. The Energy That Would Be Saved by Shifting from Each Transportation Mode to Another.[a] Btu Saved Per Dollar Saved.

Shifting from	Car	Bus	Motorcycle	Bicycle	Electric Commuter
Car		-100,000+	-392,000-	-177,000-	+17,000+
Bus	+100,000-			-51,000-	+200,000+
Motorcycle	+392,000+			-61,000-	+80,000+
Bicycle	+177,000+	+51,000+	+61,000+		+84,000+
Electric Commuter	-17,000-	-200,000-	-80,000-	-84,000-	

[a] Plus or minus signs preceding numbers indicate, respectively, an increase or a decrease in energy use. Plus or minus signs after the numbers indicate, respectively, an increase or decrease in dollar cost.
Source: Calculated from Table 10-2.

Table 10-4. Total Energy and Labor Intensity of the 20 Activities of Personal Consumption Expenditures (PCE) Highest in Terms of Dollars. Ranked in Order of Decreasing Energy Intensity, Numbers in Parenthesis are for Aggregated Activities, 1971

Activity	Energy Intensity Btu	Labor Intensity Jobs/$ Hundred Thousand
Electricity	502,500	4.4
Gasoline and oil	480,700	7.3
(Housing)	(144,000)	(N/A)
(Auto ownership)	(111,500)	(8.1)
Cleaning preparations	78,100	7.3
(Average for all personal consumption)	(70,000)	(8.0)
Kitchen and household appliances	58,700	5.5
New and used cars	55,600	7.8
Other durable house furniture	54,600	8.9
(Private investment)	(45,600)	(6.6)
Food purchases	41,100	8.5
Furniture	36,700	9.2
(Federal spending)	(36,300)	(8.2)
Women and children's clothing	33,100	10.0
Restaurants	32,400	8.8
Men and boys clothing	31,400	9.8
Religious and welfare activity	27,800	8.6
Private hospitals	26,100	17.2
Automobile repair and maintenance	23,500	4.8
Financial interests except insurance	21,500	7.8
Tobacco products	19,800	5.8
Telephone and telegraph	19,000	5.9
Rented home (interest plus some utilities)	18,300	3.5
Physicians	10,700	3.3
Own home (interest charges only)	8,300	1.7

Source: Bruce Hannon, "Energy, Growth and Altruism" (Urbana: Center for Advanced Computation, University of Illinois at Urbana-Champaign, October 21, 1975): 39.

dollar, the total amount of energy consumed would be greater than if the switch in transportation mode had not been made at all.

This example illustrates the difficulty of saving energy by changing one's life-style and purchasing habits. Individual efforts to conserve energy are further complicated by the fact that most forms of energy cannot be hoarded except by the use of energy coupon rationing. Unless energy prices rise for everyone, any reduction in consumption by one person will serve only to decrease the price of energy slightly, which will cause an increase in consumption by someone else. The complexity of the economy makes an overall increase in energy prices, relative to the prices of labor and capital, the only effective way to save energy.

NOTES

1. The available information on expenditures for each income class were translated into energy consumption in order to plot the energy curve shown in Figure 10-1. This curve includes not only the direct energy yield, but also the energy embodied in material goods and investments.

2. See U.S. Department of Commerce, Office of Business Economics, *Input-Output Structure of the U.S. Economy: 1963*, Vols. 1, 2, 3 (Washington, D.C.: U.S. Government Printing Office, 1969); and U.S. Department of Commerce, Social and Economic Statistics Administration, Bureau of Economic Analysis, *Input-Output Structure of the U.S. Economy: 1967*, Vols. 1, 2, 3 (Washington, D.C.: U.S. Government Printing Office, 1974).

3. For a detailed discussion of the relationship between employment and energy use, see Bruce Hannon, "Energy, Labor, and the Conserver Society." *Technology Review* 79, 5 (March-April 1977): 47-53.

✱ *Part IV*

Energy Production and the Community

Chapter 11

The Impacts of a Nuclear Power Plant on a Local Community: Problems in Energy Facility Development

John H. Cumberland

The major purpose of this chapter, which examines some of the problems created by the first nuclear power plant to be constructed in Maryland, is to explore opportunities for mitigating future damage to local communities from large energy facilities.

The nuclear age arrived in Maryland in 1965 when the Baltimore Gas and Electric Company announced that it intended to build a nuclear power plant at Calvert Cliffs on Chesapeake Bay. In those days there were no problems of permits or discussions with public bodies. The company had simply acquired a site from the family of a major political figure in the state, developed its plans, and was ready to proceed before it made the public announcement. Prospective construction of the Calvert Cliffs plant caused great concern because no procedures had been developed for the regulation of nuclear energy facilities by the state. As is the case with so many problems, this concern was dealt with by passing a law. The Maryland Power Plant Siting Act, adopted in 1970, was one of the first state regulatory acts for nuclear facilities in the United States. It includes some farsighted and very constructive features, although some parts of its provisions could definitely be improved.

The effects of constructing the Calvert Cliffs plant have been quite different from those predicted by the power company and by the Atomic Energy Commission. Company spokesmen asserted that construction of the plant would result in a reduction of property tax burdens, a major concern in the small rural county where it was to be located. It was also predicted that the plant would generate a large amount of local income. The Baltimore Gas and Electric Company asserted that it wanted above all to be a good neighbor. It promised to observe all environmental regulations and to cause no environmental damage. It promised that local residents would incur no impacts, except for the additional expenditures that would raise local incomes and tax revenues.

The actual reality diverged greatly from the promises. Residents learned first of all about the time lag between the construction of a plant and the period when tax revenues start flowing. During the ten years of construction, the land on which the plant was located continued to be taxed at a rate that was based on its rural farmland value; the three-quarters of a billion dollar project paid only about $40,000 per year in taxes to the county during those early years. Meanwhile, government expenditures, the county tax assessments, and bonded indebtedness were increasing sharply to accommodate the great influx of workers. The county had a population of less than 20,000 before construction of the plant began; only a very small percentage of the several thousand workers had to move into the county to exhaust all unused capacity in local government services. A housing shortage emerged in the county.[1] Schools became overcrowded. Local roads were damaged by the commuter traffic and the heavy trucks traveling into the construction site. Even though some small local businesses received increased revenues from the construction, indirectly causing a marginal increase in tax collections, there were net deficits in the county budget until the plant began operating. Property tax bills have risen, not fallen.

There were social effects as well. The admissions to the local hospital for alcoholism went up sharply. The number of traffic accidents increased greatly. There were increases in divorce rates and drug offenses. Crime became common in a county that had been essentially crime free previously. In short, a sleepy, rural area with a plantation economy suddenly was brought into the twentieth century.

There were also political impacts associated with plant construction. The Baltimore Gas and Electric Company wished to minimize the risk of having its vast investment of nearly a billion dollars questioned or delayed in any way. By allocating a fraction of 1 percent of total costs per year to public relations efforts it was able to mount a very persuasive local public relations campaign. As a first step, the company moved a resident public relations specialist into the county. By becoming president of the local Chamber of Commerce and mobilizing the local businessmen, this specialist was able to generate a great deal of convincing local support that effectively drowned out the efforts of the few volunteer environmentalists who lived in the county in those pre-earth-day and pre-sun-day years.

The company gained the early support of the then only local newspaper and took out full-page ads whenever it had an announcement to make. It also flew the editor of the newspaper, also the county's only state senator, and other influential local figures around the state in a company helicopter.

Those political figures who were not in favor of the plant, or who dared to question it, did not fare as well. People found that it was not a good idea, if they sought political or business careers in the county, to be vocal in opposing the plant or in recommending more effective regulation of it. For example, the same local newspaper gave extensive publicity to the fact that a county commissioner who had not always looked completely favorably upon the plant was holding two county offices and being paid for both. In effect, the political and economic structure of the county became dominated by the vast flow of money that is associated with the construction and operation of a nuclear power plant. Although the utility used its vast economic and public relations power to publicize items favorable to it, it also suppressed information it regarded as unfavorable. For example, when the company was required to conduct emergency evacuation tests, it attempted to conduct a trial evacuation without informing the public. Very little publicity was given to the conviction of subcontractors at the plant for accepting bribes and to the structural flaws that resulted.

The environmental impact of the plant has been devastating for the county. Billions of gallons of water are taken into the plant every day, heated, and released into an important marine spawning area of the Chesapeake Bay. Although much money has been spent to provide protective facilities, aquatic life has been damaged. A productive oyster bed was located where the company thought the plant's heated water outlet should be. The company decided that the oyster bed had to be moved, and in effect destroyed the oyster bed in order to save it. Rather than living up to all environmental regulations, as it had promised in public hearings to do, the company has lobbied tirelessly and successfully to have state environmental standards relaxed, especially on thermal releases. Most recently, the company is in extensive violation of state laws prohibiting storage of radioactive waste.

The plant's aesthetic intrusion is one of its most objectionable features. If the Baltimore Gas and Electric Company had been willing to spend a little more money, it could have screened the plant in a ravine back from the Chesapeake Bay shore rather than allowing it to spoil what used to be an unspoiled sandy beach. To reduce further the visual impact, the power lines from the plant could have been screened by trees and routed through ravines. The facility could have been built to a lower profile. Unfortunately, the plant is now a nonconforming visual intrusion over hundreds of square miles of recreational area which previously had been characterized by the unspoiled landscape and seascape that was seen by Captain John Smith when he sailed up the bay in the seventeenth century. Many of the environmentalists' objections to the plant were reactions to poor aesthetic planning.

In summary, even though modern large-scale power plants generate large amounts of energy, they do not make confortable neighbors. Careful analysis of the Maryland experience could save other regions from repeating these costly errors. Even with the best of intentions, however, it is clear that a facility of this size will cause extensive economic externalities, or effects on other parties, that are not paid for in the cost of the facility and by the sales price of the energy sold. An important fact of life about a nuclear power plant is that there is a spatial disparity between the distribution of the benefits and the distribution of the costs. The benefits go largely to affluent urban consumers, stockholders, and technicians who work on the plant. The costs are borne by those in the plant's immediate vicinity. This sort of disparity exists in almost every case where a major energy facility of any kind is constructed.

This experience and these observations of the negative effects of nuclear power plants lead to an examination of what society can and should do about this type of problem. More fundamentally, we must ask how much growth of what type is wanted, and what tradeoffs among growth, environment, and energy are to be accepted. These are very large issues with which our society has yet to deal adequately.

There are at least three types of approach that can be used to manage a situation in which extensive externalities occur. One approach is regulation; this is the typical route that American public utility commissions have adopted. Under this approach, laws are written, standards are set, and enforcement is attempted. Another major approach or technique used in the United States is the subsidy. The history of American economic development shows that, in general, we get new programs and new technologies by paying subsidies to existing interests to accept them.[2] Nuclear energy has been heavily subsidized in the United States.

The third approach, favored by most economists but distrusted by the general public and very seldom experimented with, is the use of financial economic incentives such as emissions charges. Theoretically, such charges could bring the social costs of a facility more into line with the private costs. The price of energy would rise, causing consumers to use it more economically and more efficiently, while subsidization of damage to the environment would be avoided. The Maryland Power Plant Siting Act, for example, includes an innovative energy surcharge with proceeds used for research and site acquisition.

This act could be greatly improved by converting the surcharge into a tax on emissions and by eliminating the site acquisition provision, which is simply a subsidy to an already oversubsidized industry.

How can these three regulatory devices be used to mitigate the damages and externalities from large energy facilities, including nuclear power plants?

As a regulatory effort, a moratorium on additional energy facilities ought to be considered. A noted author on this subject, Nicholas Georgescu-Roegen, suggests that there is a strong relationship between the physical science concept of entropy and economic development.[3] Georgescu-Roegen suggests that depleting our nonrenewable energy resources cannot go on indefinitely at the present rate, for use of energy increases the entropy or disorder in the world and materially reduces the amount of energy that will be available to future generations. Increases in economic welfare and development could continue even with conservation and more efficient use of current energy facilities.

The use of low-entropy energy sources could also be promoted with the second device, the subsidy. One wonders how much better off we would be were we to spend on the development of solar power and other low-entropy energy supplies a fraction of the money that is being spent on nuclear energy.

Subsidies could also be used to deal directly with the benefit-cost distribution problems created by more conventional energy facilities. For example, there have been tentative proposals to give federal and state governments power to override the wishes of local communities regarding the location of energy facilities, on the assumption that such facilities are in the national interest. The danger of such legislation could be reduced if local governments were given loans and grants to help ease the disruptions of economic and social systems brought on by large-scale energy development. The recent amendments to the Coastal Zone Management Act appear to be a step in this direction. Under these amendments, coastal communities can be provided with loans during the interim period between the construction of energy facilities and the time when the local government begins to receive increased revenue. The act also provides compensatory grants for the irremedial loss of environmental resources.

The third remedy, economic incentives, could also be applied more extensively than it is at present. This approach could help to resolve not only some of the problems faced in deciding about energy facilities but also disputes arising over the proper rate and composition of growth. For example, a three-tiered system of federal, state, and local incentives, reflecting the federal structure in this country, could be established.[4] The federal government could apply a carefully calculated emissions charge to all types of externalities—including, for example, heated water and particulate matter from fossil fuel plants—and to the generation of radioactive waste from nuclear plants. The charge would be set at a level eliminating the difference between private costs and social costs. State government would have the option of adding a surcharge to the federal emissions charge. States seeking economic development and not concerned about environmental problems could simply live with the federal charge and add nothing to it. States with a strong preference for protecting environmental quality and for slowing and changing the composition of growth could add large emissions surcharges. Local governments would also have the opportunity to add surcharges to the federal and any state charges in order to reflect local environmental and growth preferences.

This method would, to some extent, avoid the problems of letting local governments frustrate national objectives. Local communities could vary the amount of surcharge imposed both to attract facilities where they were wanted and to zone them out where they were not. Developers could offer to pay communities surcharges high enough to compensate them for damages and to overcome objections to development. On the other hand, it may appear to many people to be a subversion of the national interest to give local governments this much power. One might well ask whether it is in the national interest to give any small local community the option of putting such a high charge on the location of a needed energy facility that it cannot be built in the community.

It is argued here that it is very much in the national interest to allow such a local option. Local sovereignty is an important part of our federal political tradition. Moreover, local activist groups and local governments can often obtain environmental and other concessions from energy facility developers even when the federal government cannot. In Calvert County, Maryland, for example, the Columbia Gas Company planned to build a very large gas terminal on previously designated park land with the actual approval of both the federal and state governments. The company had ignored important environmental and aesthetic safeguards in developing its plans. Thanks to a suit by the Sierra Club, however, Columbia agreed to leave public access to a public beach, to preserve some ponds in which there were endangered species of fish and wildlife, and to screen the terminal by locating it in a place where tree cover was available.

As another example, utility company estimates of future energy needs are frequently proving to be much too high. A number of proposed nuclear and fossil fuel plants might either be delayed or canceled, with no diminution of welfare, if local governments and citizens' groups are able to mobilize to bring such changes about. Where there is strong local opposition to an energy facility, an effective demonstration of this opposition is very much in the public interest.[5]   Effective local resistance to large energy technologies can also provide strong incentives to develop less damaging technological advances.

Giving states and communities an opportunity to express local preferences and concerns for the balance between growth and environment by imposing local emission charges could do more than mitigate the impacts of energy facility development. It could simultaneously help solve the critical current problems of the appropriate rate of growth, the composition of growth, and the spatial distribution of growth.

NOTES

1. John H. Baldwin *et al.*, "Socio-Economic Impact of Power Plant Construction: A Case History," in *Record of the Maryland Power Plant Siting Act*, vol. 4, no. 3 (June, 1975): 1-6.

2. John H. Cumberland, *Regional Development Experiences and Prospects in the United States of America* (Paris: United Nations Research Institute for Social Development, 1971).

3. Nicholas Georgescu-Roegen, "Energy and Economic Myths," *Southern Economic Journal*, 41, 3 (January, 1975): 347-381.

4. John H. Cumberland, "Boundary Conditions and Influence on the Planning of the Power Generating Industries," in *Energy and Environment (Energie und Umweldt)* (Essen, West Germany: Vulcan Verlag, June, 1977): 43-45.

5. John H. Cumberland, William Donnelly, Charles S. Gibson, Jr., and Charles E. Olson, "Forecasting Alternative Electric Requirements and Environmental Impacts for Maryland, 1970-1990" in *Energy, Regional Science, and Public Policy*, edited by M. Chatterji and P. Van Rompuy (New York: Springer Verlag, 1976): 32-57.

Chapter 12

Boom or Bust:
Energy Development in the
Western United States

John S. Gilmore

This chapter describes some of the problems that have arisen with the building of power plants in isolated rural communities in the Rocky Mountains-Great Plains region. Most of the affected towns are farming and ranching communities with some mining activity. Although two communities in Wyoming will serve as examples, the same difficulties have been experienced in towns in Utah, Montana, Colorado, and Alberta. Some possible solutions to these problems will also be proposed.

Rock Springs and Green River, Wyoming, two tiny mining towns in huge, sparsely populated Sweetwater County, were faced in 1972 with several startling announcements. All of the large mining companies operating in the county declared that they were planning immediate, major expansions with large increases in employment. In addition, Pacific Power and Light Company announced that it was going to build a 2,000-megawatt, coal-fired steam power plant in the area. It estimated that as many as 1,200 construction workers would have to be brought into the county by 1975 to complete the project.

The dramatic growth brought on by these expansions was not easily absorbed by the towns. Such isolated communities can generally accommodate a 5 percent annual growth rate with no major problems. Construction of new housing and of public facilities are able to keep pace. With growth between 5 and 10 percent, however, delays begin to occur, and when the growth rate rises above 10 percent, serious problems of school overcrowding and housing shortages become widespread. Rock Springs and Green River were growing at a rate of greater than 15 percent, fast enough to cause real institutional breakdowns. As the towns became more overcrowded, productivity went down, and more workers had to be brought in, exacerbating the existing situation. By 1973, for example, 3,000 people were employed in construction of the power plant. The population of the area doubled within three and one-half years after the boom began.

The decrease in the quality of life in Rock Springs and Green River was most apparent in the housing problems the area experienced. The housing shortage caused by the boom was so severe that 75 percent of the incremental population had to be housed in mobile homes. Because there was no subdivision regulation, zoning, or any other type of land use control, the trailers were literally strewn about on the prairie. Unsanitary conditions were common in the mobile home "communities." Water frequently had to be hauled in, and it was common to see children playing in pools of sewage from malfunctioning septic tanks. Those in trailers were relatively well off, however, compared to the workers who had to live in tents. The few houses available for purchase cost so much that scarcely anyone could afford the mortgage payments. No public housing was available. Housing prices rose so high that many retired, older residents of the county had to move away.

To make matters worse, the local financial institutions were not interested in financing housing of any type for anyone they did not know. They seemed to hope that the boom would simply go away. It was inconvenient to them, and they wanted no part of it.

Although the housing shortage was the major problem, health services also were grossly inadequate. There was a decrease during the boom period in the number of practicing physicians. With only one doctor per 3,700 people (as opposed to one per 1,000 before the boom), it took two to three weeks to get an appointment. There was no alternative except going to the hospital emergency room if one needed to see a doctor quickly. The number of emergency room admissions averaged 2,000 per month despite a minimum charge of $26.

During the first three and one-half years of the boom, mental health caseloads increased about 900 percent. Many of the new clients were workers' wives, who had nothing to do during lonely days in the isolated trailer camps except listen to their babies cry or the wind blow, and wonder if their husbands would come home sober. Few jobs other than mining and manufacturing were available for wives who wanted to work, and no social or recreational activities were available. (Fortunately, the local junior college has since begun to provide day care and transportation in conjunction with new programs geared to workers' wives. This should relieve some of the anxieties that these women face.)

Local educational and recreational facilities were badly strained by the boom. The school districts were forced to go beyond their legal, constitutional bonding limits in their attempt to provide for the increased demand. Private recreational facilities became unpleasant places to go, because of the many fights that occurred. Families became afraid to go out at night, leaving the streets and restaurants to the construction workers.

There was a rapid increase in the crime rates in the county, and new criminal activities, such as prostitution, were introduced to the area. A dislike for the changes that were occurring in the character of the community caused many citizens to perceive the increase in crime to be even greater than it was.

As noted above, a reduction in industrial productivity and reliability accompanied the decline in the quality of life. Increases in turnover rates and absenteeism were common. Productivity in existing mines, measured by how much was produced per man per shift, dropped 25 to 40 percent within a twelve-month period. A similar decrease occurred on the power plant construction site, although the company was reluctant to release exact figures. The tremendous size of the power company's productivity loss was indicated by the fact that the final cost of the plant was 65 to 100 percent more than expected, and by the increases, mentioned above, in the number of employees needed to build the facility.

The delays experienced by the power plant caused a delay in the increase in local tax revenue that the completed plant was to provide. This further diminished the already threatened fiscal viability of local government. Assessed valuation per capita, measured in real dollars, dropped in three years to a little less than half of what it had been before the rapid growth started, while the need for local services increased rapidly, as noted above. Thus, a self-perpetuating cycle was created. As the local services fell short of the needs of the increasing population, the quality of life declined, causing increases in turnover rates, absenteeism, and other industrial employment problems, and decreases in productivity and profitability. Industry was unable to increase output, and private investors were reluctant to provide the services the growing population needed. Little growth in output and investment meant smaller increases in tax receipts and a further crippling of local governments' ability to keep up with greatly increased demand.

A number of problems and issues are raised by rapid growth situations such as the one described. They include:

1. Financing housing and public facilities, given the breakdown of the capital market. It is difficult even to attract developers to an area if there is danger that they will lose their skilled laborers to the local heavy industries.

2. Controlling fringe settlements in an area where police power regulations, such as zoning, are politically impossible to enact.
3. Keeping high quality employees in local government when it cannot afford to pay salaries as high as those in industry.
4. Determining the proper mix of public, private, and voluntary institutional activities needed to make life tolerable for families.
5. Determining the proper governmental structure for dealing with boom town impacts. Persons working on energy projects often live in a different town or county from the one in which the facility is located, so that the government receiving tax revenues from the project is often not the one faced with providing the extra services.
6. Accommodating new growth when a limited amount of land is available. Towns in the West are often constrained in the amount they can expand by the fact that the federal government and private companies own large portions of the surrounding land.
7. Meeting the challenge of urban problems by means of transition to a more urban form of government. In many of the boom towns, defeat of incumbent officials is common as citizens become impatient with those officials' inability to handle the complex new problems created by rapid growth.
8. Achieving cooperation between industry and government in dealing with boom town problems, while avoiding industry domination and preventing industry apathy.
9. Dealing with a specialized economy very susceptible to "crashing" or "busting" because of its dependence on a depletable resource.
10. Handling the breakdown of institutions that often accompanies very high growth rates.

At least four possible solutions to the problems of boom town development can be suggested. First, investment must be balanced between basic industry and local services. About 5 to 20 percent (depending on the labor intensity of the basic industry) of the amount spent on industrial capital should be spent on providing the service infrastructure. The major problem encountered in achieving such a balance is the lack of communication between basic industry and the public and private service sector investors. This difficulty is complicated by the fact that the service investors may not be able to obtain the capital they need until long after a basic industry investment decision is made.

To deal with this problem in Sweetwater County, A Sweetwater Priorities Board was established, consisting of eight elected officials, five representatives of industry, and two citizens-at-large. The board's first task was to deal with the crisis situation existing in the county. It had to find funds where none were available to solve the county's most pressing problems. Later, citizens' task forces were set up to work with individuals on the board on particular issues. The board's activities are monitored closely at present by the League of Women Voters and other community groups. Thus far, it seems to be working fairly well.

The Federal Power Commission also has been involved in attempts to balance basic and service investment. A few years ago, the Virginia Electric Power Company was granted a conditional permit by the FPC to construct a power plant, provided that it supply the needed housing and other facilities for the growth its project would cause.

This same sort of negotiation recently occurred in Wheatland, Wyoming. Before Basin Electric Company was given a state permit to construct a 1,600-megawatt plant in that town of 3,000, it was required to commit $22 million for service infrastructure. It also had to agree that during the course of construction it would do almost anything requested by a locally dominated group set up to monitor the plant's impact.

A second way to solve or prevent boom town problems is to regulate resource use and conservation. Because the conventional technique for such regulation, zoning, is not popular in the Rocky Mountain and Great Plains areas, some states, such as Wyoming and Montana, have passed rigorous plant siting legislation as an alternative strategy. Such legislation preempts local prerogatives regarding siting questions, in an attempt to bring some order to land use decisions for energy projects.

Development of the existing local labor force is a third way of easing rapid growth problems. If more people in the present population are hired, fewer new workers from outside the area need to be recruited, and less new housing is needed. For example, a coal-mining firm in southwest Colorado has hired, almost exclusively, inexperienced local people, both men and women, and trained them on the job. The company reported that productivity is extremely high.

A fourth way to alleviate some of the problems associated with energy development projects is to provide the necessary human support services to accommodate and retain the existing population. Both formal public programs and efforts by voluntary associations are needed. State governments and industry should use any discretionary money they have available to help increase the effectiveness of volunteer groups' efforts.

Several questions about the solutions to boom town problems remain unanswered. First, under what conditions is it feasible to pro-

vide loans or loan guarantees to develop the local service infrastructure to balance basic investments? Such loans are often politically
unpopular, for early debt service requirements result in an increase
in taxes until the new industry is in operation and the tax base
expands.

Second, what are the proper roles of the federal and state government in dealing with boom town problems? More importantly, how
do these roles fit together?

Finally, what does it really mean to "mitigate" the socioeconomic
impacts of energy development projects? Does one avoid the problems altogether, or does one alleviate them once they occur? Which
technique is most effective? Which is most expensive? What risks are
involved in choosing one alternative instead of the other?[1]

NOTES

1. Readers interested in related reports of Gilmore's research may want to
consult John S. Gilmore, "Boom Towns May Hinder Energy Resource Development," *Science*, 191, 4227 (February 13, 1976), pp. 535-570.

✳ *Part V*

Research Needs in Energy and the Community

Chapter 13

Difficulties in Assessing
the Socioeconomic Impacts
of Energy Facilities

Stephen S. Skjei

The Nuclear Regulatory Commission (NRC) derives its authority primarily from the Atomic Energy Act of 1954 and the Energy Reorganization Act of 1974. Under the Atomic Energy Act, the major responsibility of the NRC is to protect the health and safety of the populace in the development and use of atomic energy for peaceful purposes. The Energy Reorganization Act divided the Atomic Energy Commission into the NRC and the Energy Research and Development Administration. The NRC is also governed by the National Environmental Policy Act. That legislation, and court decisions interpreting it, require the NRC to make assessments of both the environmental and socioeconomic impacts of nuclear generating stations.

The socioeconomic problems caused by most nuclear generating stations are significantly different from the difficulties associated with isolated energy developments in the western United States as described by John Gilmore. Three dimensions of the socioeconomic impacts of nuclear generating stations are of particular note: labor force, location, and tax base.

Currently an average of 700, and a peak of 1,700, construction workers may be hired to build a one-unit nuclear generating station. The number of workers needed to construct a two-unit nuclear generating station averages about 1,500, with the peak on-site employment for such a station being about 2,800. Construction employees are on the site for three to six years per unit. They create most of the short-term socioeconomic impacts commonly associated with nuclear generating stations. By contrast, the operating labor force is small, ranging from 150 to 250 workers, depending on station size. This labor force is not large enough to cause a significant socioeconomic impact on a community.

The location of a nuclear generating station is as important as labor force characteristics in determining socioeconomic impacts. Most stations are not built in isolated areas. Presently, about thirty-eight nuclear generating stations are under construction or in operation less than 30 miles from the central city of a Standard Metroplitan Statistical Area (SMSA), whereas forty-two are between 31 and 60 miles away. Only thirteen are being constructed or are in operation more than 60 miles from the central city of an SMSA, and all of these are within 100 miles of such a city. As an example, eight nuclear generating units are presently in operation with 75 miles of downtown Chicago and another seven are under construction within that same radius.

Proximity to an urban area makes a difference in the type of socio-economic impacts that occur. When laborers can commute from existing urban areas to the construction site, not as many people move to the community where the site is located, and local facilities are generally not overburdened. Overuse of local roads is the major problem that occurs.

The third aspect of nuclear generating station impacts is tax base changes. In many cases, nuclear generating stations provide substantial tax revenues to the communities in which they are located. Plymouth, Massachusetts, and Waterford, Connecticut, are good examples. As a consequence of the construction of nuclear generating stations, the tax base of Plymouth doubed; that of Waterford increased by 33 percent. Despite these impacts, the view has been expressed more than once that communities in which nuclear generating stations are being constructed should be compensated for the dislocations they endure and the expenses they may incur as a consequence of construction.

In some states legislation has been passed that requires that the tax revenue generated by a nuclear generating station be shared among all the state's communities. Locating new stations in states with such legislation can be difficult because communities object to absorbing the external costs of a station when they will receive only a fraction of the tax revenues.

Assessing the socioeconomic impacts of a nuclear generating station located near a growing urhan area is very difficult. The impacts associated with plant construction often cannot be distinguished from the impacts occurring naturally because of the growth of the region in which the station is located. Conceptually, three broad classes of impact should be assessed: infrastructure, service delivery, and quality of life. Infrastructure impacts include those on schools, roads, the sanitation system and the sewer system. The effects of power plants on the health system and on police and fire protection are examples of service delivery impacts. Effects on housing, taxes, future development, regional income, noise levels, and aesthetics can be included in the category of quality of life impacts.

The researchable question is whether and to what extent these impacts are associated with the construction of nuclear generating stations. Answering this question can be very complex. An ex post study of nuclear generating stations in Plymouth, Massachusetts, and Waterford, Connecticut—both areas presently undergoing urban growth—could, with the exception of tax and traffic impacts, find no clearly distinguishable socioeconomic impacts attributable to either of these energy facilities.[1] Perhaps if thirty or forty plants had been studied, a statistical analysis could have been performed and a relationship between nuclear generating stations and socioeconomic impacts could have been found. Such a study would be both difficult and expensive to complete, however.

To date, most analyses of the socioeconomic impacts of nuclear generating stations have been ad hoc. Usually, the researcher observes a community with a nuclear generating station, makes note of the impacts that he thinks he sees, and develops his findings. Such an approach usually provides no clearcut evidence that what has been observed can be attributed to the nuclear generating station. Also, without a theory linking socioeconomic impacts to large energy developments, the data collected cannot conceptually be interpreted in a meaningful manner.

An interesting question is what kind of theory or set of hypotheses is needed to relate observed socioeconomic impacts to the construction of a nuclear generating station. Does any applicable theory exist? Standard techniques for regional analysis do not provide an appropriate conceptual base for asserting any connection. Journey-to-work hypotheses might be useful, but most are applicable only within an urban area. They are not useful in analyzing the travel of construction labor from the city to a plant site in the surrounding countryside. A modified Tiebout hypothesis also might be used to study the impacts on urban development of the increased tax revenues and/or lowered tax rate resulting from the location of a nuclear generating station in a community. However, substantive data manipulation would be required to obtain a meaningful result with this hypothesis. In short, a problem described long ago by Andreas Papandreou in his discussion of "theorems of applicability" is encountered in evaluating the socioeconomic impacts of nuclear generating stations. The theorems and constructs that have been used to look at the real world to determine what is happening and to guide research do not appear to be applicable to this type of analysis. To make matters worse, even less theory is available on which to base predictions of the impacts of facilities not yet constructed. Because of this lack of theoretical knowledge, policy and program analysts are forced to rely heavily on descriptive techniques in making their assessments of the socioeconomic impacts of nuclear generating stations. Even though this might ultimately result in the development of a conceptual basis for analyzing these socioeconomic impacts, it can do so only if the accuracy of predictions is assessed, and to date this has not occurred on a widespread basis.

The scarcity of data also complicates the assessment of the socioeconomic impacts of nuclear generating stations. Although some information about impacts may be obtained through surveys of people in the communities affected, such research cannot substitute for reliable, factual data on the impacts that are of concern. Unfortunately, the experiences of the researchers at Plymouth and Waterford indicate that such data are hard to obtain.

The combined lack of theoretical and empirical knowledge about energy facility impacts complicates our understanding of them. Until standardized procedures can be developed to compare the effects of a large number of facilities, we will continue to have a very fragmented picture of these impacts.

NOTES

1. B. J. Purdy, et al., "A Post Licensing Case Study of Community Effects at Two Operating Nuclear Power Plants." Draft (Oak Ridge, Tenn.: Oak Ridge National Laboratory, 1976).

Chapter 14

Federal Research Programs
for Energy Conservation
in Buildings and Communities

Kim Gillan

The National Plan for Energy Research, Development, and Demonstration lists a number of reasons why conservation should have high priority as an energy strategy for the nation. First, a reduction in all consumption of oil will reduce petroleum imports and the nation's dependence on foreign supplies. Second, it typically costs less to save energy than to produce more energy by the development of a new supply technology or the construction of a new facility. Third, conservation techniques generally can be implemented faster, with less government involvement, than can some of the suggested supply strategies. Fourth, energy conservation also provides environmental benefits.

The Energy Research and Development Administration's (ERDA's) Office of Conservation has concentrated its research efforts on programs of energy efficiency and fuel switching as important ways of helping the nation meet future energy demands. Increased efficiency means the production of goods and the provision of services with the use of less energy, sometimes at a lower total cost. Fuel switching involves the substitution of more plentiful energy sources for oil and natural gas. Most of the office's research programs are not oriented toward developing strategies that will require changes in people's behavior or life-styles. Instead, ways are being sought to provide the same services with less energy. Research in the area of lighting is an example. Rather than exploring methods for persuading people to turn off their lights, research support has been provided for the development of a fluorescent screw-in light bulb that uses only half the energy of a traditional bulb.

It is often asked why government has a role in energy research and development and in energy conservation. Some people think that these should be primarily private sector activities. They argue that if the economic situation really justifies conservation, private companies would already be pursuing research and development activities in that area, for they would see that conservation technologies are marketable.

At the present time, however, energy prices are artifically low. As a result, some energy conservation technologies are not yet very appealing. One has to consider the energy savings over a number of years in order to rationalize the high initial cost of many energy-saving devices. No regulatory systems exist to encourage energy conservation.

In such a situation, the federal government can complement the research, development, and marketing efforts of the private sector. The Office of Conservation provides funds to develop promising new technologies that private industry still considers too risky. It aids in the introduction of products of which the user is wary, because of a lack of reliable information, and it encourages private industry in the development of markets for new energy-saving items.

The development of the fluorescent light bulb mentioned above provides an example of the way in which the federal government can encourage the research efforts of private industry. When ERDA was first considering funding research for the development of more energy-efficient light bulbs, some of the traditional light bulb manufacturers argued that energy prices would not justify work on such products for five or ten more years. Initially, ERDA funded an individual in California to do the conceptual development of such a bulb. The challenge of competition then made the other manufacturers speed up their own research activity in this area.

The fluorescent light bulb that has been developed has a life of thirty years, but it costs about $10, more than most people will pay with today's energy prices. It is expected, however, that as energy costs increase, this bulb will be seen as more efficient and economical than today's model. The Office of Conservation makes the stipulation concerning all of its research projects that any technologies that are developed must be capable of being accepted into the market within a reasonable period of time. ERDA's first priorities are not on strategies requiring heavy subsidies to be implemented on a widespread basis.

The research goal of the Division of Buildings and Community Systems is in keeping with the Office of Conservation's objective of promoting energy efficiency. The division is charged with developing technical, financial, and institutional strategies to promote the use in buildings, communities, institutions, and homes of more energy-efficient technologies.

There are four sections with separate research programs within this division: consumer products and appliances; architecture and engineering; urban waste; and community systems. The consumer products and appliances section looks for ways to make home heating and lighting systems, refrigerators, and similar items more energy efficient. It also examines building controls for smaller buildings.

The researchers in the architecture and engineering section study the energy utilization of residential, commercial, and institutional buildings. They are advising the Department of Housing and Urban Development concerning energy efficiency standards for new buildings, and they are working with the national building code organizations to develop training manuals and model energy-efficient building codes. This group, along with the Southern California Gas Company and several other sponsors, has been involved in the development of minimum-energy dwelling units in southern California. All of the existing technologies that can be bought "off the shelf" have been incorporated in the two houses built. The researchers have examined the practicality of living in such homes, have estimated the houses' cost and marketability, and have evaluated the effectiveness of the various energy conservation techniques used.

The third group of researchers is studying the potential for using urban waste as a source of energy. Although there are many techniques and systems that can be used to transform waste into a fuel, a number of economic and institutional problems must be overcome. These difficulties are being confronted by researchers in pilot projects in Pompano Beach, Florida; Seattle; Washington; and other cities. They hope to determine what percentages of energy demand can be met by a system using a mixture of urban waste and conventional fuels.

The fourth research section, community systems, is involved in research concerning the spatial relationships of buildings in communities. Transportation, street lighting, and utilities are also studied in an effort to see how communities use energy, and how that use might be made more efficient. The remainder of this article describes the federally supported research being undertaken on energy and the community.

Integrated community energy systems are one major research interest. Traditionally, systems using energy in different forms (oil, natural gas, and electricity, for example) have been thought of as separate units, all with their own inefficiencies. Recently, however, the idea of combining the production of energy of different types in the same system has been suggested as a way of increasing overall efficiency. Pilot programs in on-site electric power generation with reuse of waste heat are currently under way on five sites in the United States.

Researchers are also trying to determine whether integrated community utility systems and other new technologies will be compatible with existing patterns of growth. On-site power plants, for example, require certain demands at certain times to be efficient, and land use plays a large role in determining the shape of energy demand. As another example, district heating and cooling techniques, long employed in Sweden and Germany, need clustered, high-density development with residents close to industry to be effective, whereas individual solar collectors require a relatively large amount of space for each unit. We need to know more about the locational requirements of new energy technologies in order to take them into account in designing land use and comprehensive plans.

Research on community design is also being pursued. Investigators are trying to determine what impacts design has on community energy use, and to decide how design might be used as a tool for achieving greater energy efficiency. The way a community is designed is the result of many decisions in both the public and private sectors. Decisions regarding land use, transportation, water, sewer, and educational facilities, among others, all affect development at the site, in the neighborhood, and in the community.

There is much debate about how effective or efficient land use planning for energy conservation can be, compared with other strategies that are being developed that can guarantee definite reductions in energy consumption. It is not at all clear how much energy can be saved by means of integration of activities, by means of clustering of development, or by means of changes in transportation.

Instead of attacking this problem by developing impractical conceptual designs of very energy-efficient communities, a process-oriented approach is being pursued. The focus of this approach is on the development of analytical tools and procedures for energy conservation planning that can be used in the continuing decision-making process of community design. This is a difficult area in which to work, for one is never dealing with tangible products.

The community design research program has four parts. First, as noted above, methodologies and analytical tools that communities need to do energy planning are being developed. Energy audits and benefit-cost analyses are examples of techniques that are being evaluated. The goal of these efforts is to make energy considerations a part of existing planning and decision-making processes. Once a generic energy planning methodology is developed, communities will be invited to submit proposals for using the specified techniques to develop community energy plans. In this way, the effectiveness of the methodology will be tested, and an indication of the sorts of organization that local governments need to set up to handle this new function will be provided.

Energy master planning is being explored, using several actual site development projects as case studies. Developers funded to create energy-efficient site plans are required to evaluate how much more expensive their designs are than conventional plans. They also must assess the marketability of their schemes, and determine what implementation problems, if any, are posed by zoning, subdivision regulations, and building codes. The objective of this research is to pinpoint the stages in the development process at which energy-conserving techniques can best be introduced. Developers can then be shown how they might make energy efficiency one of their objectives, independent of any type of regulation.

Another part of this energy master planning research is concerned with the design and development of Alaska's proposed new capital. Funds have been provided so that local officials can identify energy conservation alternatives for the new community that is being planned. In addition to describing the conservation methods that may actually be used, they are developing a general inventory and benefit-cost analysis of other strategies that were considered, but rejected. These alternative techniques, even though inappropriate for Alaska, may be useful in other parts of the United States.

The community design research effort is attempting to identify and substantiate the relationship between energy and land use. Much of the information presently available is speculative, because of the inadequacy of the data base from which conclusions must be drawn. Research is being funded to determine what significant energy-land use relationships exist, and to assess the energy requirements of different types and patterns of land uses. This will help in the establishment of a solid base of energy information that can be used in conjunction with existing methodologies for land use planning and decision-making.

Program support, or collection and dissemination of information, is the final part of the community design research effort. Unfortunately, because most of the research is just beginning, there is little information on energy conservation and community design yet available. In the interim, the Federal Energy Administration (FEA) has filled this gap by holding training workshops to teach communities how to do comprehensive energy planning, and working with the National League of Cities, the National Association of County Officials, and the United States Chamber of Commerce in developing information handbooks.

Even though the FEA has supplied conservation information to communities, programs have not been available to provide financial help in the development or implementation of local energy conservation plans. This is a serious hindrance in convincing local governments to implement conservation strategies. Rather than trying to solve this problem by inventing a new funding program, efforts are underway to convince other federal agencies, which already have assistance programs affecting community development, to make energy considerations part of their funding requirements. (Such agencies include the Department of Transportation, the Environmental Protection Agency, and the Economic Development Administration.) At present, some federal loan programs have energy conservation stipulations, and the federal public works program requires an energy assessment in submitted project proposals.

In the future, federal research managers hope to work closely with local public officials and professional planners, as well as with other federal agencies, in order to make research programs as useful as possible. An urban consortium of twenty-seven cities, organized through Public Technology, Inc., will serve as a "users' group" to evaluate some ongoing research projects. They will indicate how relevant ERDA-sponsored research is for existing cities. The Urban Land Institute will also assess some of these research projects. It will provide information about trends in the development industry and in planning, to help keep research in line with trends in the "real world." Research will be undertaken in conjunction with planners to develop model regulations for promoting land use patterns indicated by research as being most energy efficient. With President Carter's strong emphasis on conserving energy, the community energy conservation research program should play an increasingly important role in developing the nation's energy strategy.

Selected References on
Energy and the Community

Acton, Jan Paul, and Mowill, Ragnhild Sohlberg. *Conserving Electricity by Ordinance: A Statistical Analysis.* Santa Monica, Calif.: Rand Corporation, R-1650-FEA, February 1975.

Adler, Thomas J., and Ben-Akiva, Mosha, "Joint-Choice Model for Frequency, Destination, and Travel Mode Shopping Trips." *Transportation Research Record* no. 569 (1976): 136-150.

Allen, Edward H. *Handbook for Energy Policy for Local Governments.* Lexington, Mass.: D. C. Heath and Company, Lexington Books, 1975.

Alternatives in Energy Conservation: The Use of Earth Covered Buildings. Proceedings and Notes of a Conference Held in Fort Worth, Texas, July 9-12, 1975, Sponsored by University of Texas at Arlington. Washington, D.C.: U.S. Government Printing Office, 1976.

Andersen, Stephen O. "State Strategies for Energy Conservation by Public Utilities and Electricity Consumers." In The Conservation Foundation, *Energy Conservation Training Institute.* Washington, D.C.: The Conservation Foundation, n.d., pp. IV-185 - IV-214.

Athern, William; Doctor, Ronald; Harris, William; Lipson, Albert; Morris, Deane; and Nehring, Richard. *Energy Alternatives for California: Paths to the Future.* Santa Monica, Calif.: Rand Corporation, R-1793—CSA/RF, 1975.

Bacon, Edmund N. "Energy and Land Use," *Urban Land* (July-August, 1973): 13-16.

Baldwin, John H.; Needles, Howard, Tammen, and Bergendoff, "Socio-Economic Impact of Power Plant Construction: A Case History." In *Record of the Maryland Power Plant Siting Act*, vol. 4, no. 3 (June 1975).

Baldwin, Pamela L., and Baldwin, Malcolm F. *Onshore Planning for Offshore Oil.* Washington, D.C.: The Conservation Foundation, January 1975.

Bammi, Deepak, and Bammi, Dalip. "Land Use Planning: An Optimizing Model," *OMEGA, The International Journal of Management Science*, 3 (1975): 583-594.

Bander, Jeff; Bergheim, Mel; Hamilton, Tara; King, Norman; and Wald, Sarah. *Energy Conservation in Buildings: New Roles for Cities and Citizen Groups.* Washington, D.C.: National League of Cities and U.S. Conference of Mayors, January 1975.

Booz, Allen & Hamilton, Inc., Development Research Division. *Interaction of Land Use Patterns and Residential Energy Conservation.* FEA Task Order CO-04-50250-00, Job 13392-005-001. Report prepared for the Federal Energy Administration. Bethesda, Md.: Booz, Allen & Hamilton, Inc., October 20, 1976.

Bossong, Ken. "Organizing for Effective Citizen Action." In The Conservation Foundation, *Energy Conservating Training Institute.* Washington, D.C.: The Conservation Foundation, n.d., pp. V-37 - V-72.

Brannon, Gerard M. "Tax Policy and Energy Conservation." In The Conservation Foundation, *Energy Conservation Training Institute*, Washington, D.C.: The Conservation Foundation, pp. 111-107 - 111-131.

Brewer, William A. "State Energy Policies for the Northwest," *Washington Public Policy Notes*, 3 (July 1975).

Burley, Robert A. "Urban Form and Energy Conservation." Statement to the Subcommittee on the City, Committee on Banking, Finance, and Urban Affairs, United States House of Representatives, Washington, D.C., September 14, 1977.

Calderon, Cinda Martin, and McKenna, David. *Energy and Local Government.* Arlington: Institute of Urban Studies, University of Texas at Arlington, 1974.

Caldwell, Lynton K. "Energy and Environment: The Bases for Public Policies." *The Annals of the American Academy*, 410 (November 1973): 127-138.

Carroll, T. Owen; Beltrami, E.; Kydes, A.; Nathans, R.; and Palmedo, P. F. *Land Use and Energy Utilization: Final Report.* Springfield, Va.: National Technical Information Service, BNL 50635, June 1977.

Carroll, T. Owen; Kydes, A.; and Sanborn, J. "Land Use-Energy Simulation Model: A Computer-Based Model for Exploring Land Use and Energy Relationships." (Draft) Upton and Stony Brook, N.Y.: National Center for Analysis of Energy Systems, Brookhaven National Laboratory, and The Institute for Energy Research, State University of New York at Stony Brook, September 24, 1976.

Carroll, T. Owen; Nathans, R.; Palmedo, P. F.; and Stern R. *The Planner's Energy Workbook: A User's Manual for Exploring Land Use and Energy Utilization Relationships.* Upton, N.Y.: Policy Analysis Division, National Center for Analyzing Energy Systems, Brookhaven National Laboratory, October 1976.

Clark, James W. *Assessing the Relationships Between Urban Form and Travel Requirements: A Literature Review and Conceptual Framework.* Seattle: Urban Transportation Program, University of Washington, August 1970.

Clark, Wilson. "Innovations in Energy Conservation: Technological Applications." In the Conservation Foundation, *Energy Conservation Training Institute.* Washington, D.C.: The Conservation Foundation, n.d., pp. IV-375 - IV-433.

Committee on Investment Consequences of Urban Growth Trends. *Urban Trends and the Energy Situation.* New York: The Conference Board, June 18, 1974.

Conklin & Rosant and Flack & Kurtz. *Reading the Energy Meter on Development.* Washington, D.C.: Federal Energy Administration, November 1976.

Council of State Governments. *Energy Conservation: Policy Considerations for the States.* State Environmental Issues Series. Lexington, Ky.: The Council of State Governments, November, 1976.

Cumberland, John H. "Boundary Conditions and Influence on the Planning of the Power Generating Industries. In *Energy and Environment (Energie und Umweldt).* Essen, West German: Vulcan Verlag, June 1977.

Cumberland, John H. "Forecasting Electric Energy Requirements and Environmental Impacts for Maryland, 1970-1990." Paper prepared for presentation at the Twenty-second North American Meetings, Regional Science Association, Cambridge, Mass., November 8, 1975.

Cumberland, John H. "Interdependence Between Energy, Environment, Regional Development, and Economic Growth." Paper prepared for the Northeastern Regional Science Association, Amherst, Mass., April 18–19, 1975, revised June 1975.

Cumberland, John H. *Regional Development Experiences and Prospects in the United States of America.* Paris: United Nations Research Institute for Social Development, 1971.

Cumberland, John H.; Donnelly, William; Gibson, Jr., Charles S.; and Olson, Charles E. "Forcasting Alternative Electric Requirements and Environmental Impacts for Maryland, 1970-1990." In M. Chatterji and P. Van Rompuy, eds., *Energy, Regional Science, and Public Policy.* New York: Springer Verlag, 1976.

Dantzig, George B., and Saaty, Thomas L. *Compact City.* San Francisco: W. H. Freeman and Co., 1973.

Darmstadter, Joel. *Conserving Energy: Prospects and Opportunities in the New York Region.* Baltimore: The Johns Hopkins University Press, 1975.

Dendrinos, Demitrios S. "Energy Costs, Transportation and Urban Form." Lawrence: Institute for Social and Environmental Studies, The University of Kansas. Revised November 1, 1977.

Edwards, Jerry L. "Relationship between Transportation Energy Consumption and Urban Spatial Structure." Unpublished Ph.D. dissertation, Northwestern University, 1975.

Edwards, Jerry L. "Use of a Lowry-Type Spatial Allocation Model in an Urban Transportation Energy Study." *Transportation Research,* 11, 2 (April 1977): 117-126.

Edwards, Jerry L., and Schofer, Joseph L. "Relationships between Transportation Energy Consumption and Urban Structure: Results of Simulation Studies." Paper sponsored by Committee on Transportation Systems Design, n.d.

"Energy Conservation and Land Development," *Environmental Comment,* July 1977, entire issue.

"Energy Conservation in New Building Design: An Impact Assessment of ASHRAE Standard 90-75, Executive Summary." Paper prepared for the Office of Buildings Programs, Energy Conservation and Environment, Federal Energy Administration (Conservation Paper 43A). Washington, D.C.: Federal Energy Administration, n.d.

"Energy Policy and the Poor." *Focus,* 5, 5 (May, 1977): 1-5.

Energy Policy Project of the Ford Foundation. *A Time to Choose: America's Energy Future.* Cambridge, Mass.: Ballinger Publishing Co., 1974.

Energy Research and Development Administration. *A National Plan for*

Energy Research, Development & Demonstration: Creating Energy Choices for the Future 1976. Vols. 1 and 2. Washington, D.C.: U.S. Government Printing Office, 1976.

Energy Research and Development Administration, Office of Conservation, Division of Buildings and Community Systems. *Buildings and Community Systems Detailed Program Plan.* Washington, D.C.: Energy Research and Development Administration, September 30, 1976.

Executive Office of the President, Energy Policy and Planning. *The National Energy Plan Summary of Public Participation.* Washington, D.C.: U.S. Government Printing Office, 1977.

Federal Energy Administration. "Analysis of Thermal Standards for Residential and Commercial Buildings." Washington, D.C.: Federal Energy Administration, 1975.

Federal Energy Administration. *Project Independence, Residential and Commercial Energy Use Patterns 1970-1990.* Washington, D.C.: U.S. Government Printing Office, November 1974.

Federal Energy Administration, National Energy Information Center. *Directory of State Government Energy-Related Agencies.* Washington, D.C.: U.S. Government Printing Office, September, 1975.

Federal Energy Administration, Office of Data and Analysis, Policy and Analysis. *Energy Information in the Federal Government: A Directory of Energy Sources Identified by the Interagency Task Force on Energy Information.* Washington: National Energy Information Center, Federal Energy Administration, 1976.

Florida Energy Committee. *Energy: Policy and Recommendations for Florida.* Tallahassee: Florida Energy Committee, 1975.

Florida Energy Office. *A Planner's Handbook on Energy (With Emphasis on Residential Uses).* Tallahassee: Florida Energy Office, January 9, 1976.

Foell, W. K.; Mitchel, J. W.; and Pappas, J. C. *The Wisconsin Regional Energy Model: A Systems Approach to Regional Energy Analysis,* IES Report 56. Madison: Institute for Environmental Studies, University of Wisconsin, September 1975.

Fraker, Harrison, and Schorske, Elizabeth. *Energy Husbandry in Housing: An Analysis of the Development Process in Residential Community.* Report 5. Princeton, N.J.: Center for Environmental Studies, Princeton University, December 1973.

Franklin, Herbert M. "Will the New Consciousness of Energy and Environment Create an Imploding Metropolis?" *AIA Journal* (August 1974): 28-36.

Georgescu-Roegen, Nicholas. "Energy and Economic Myths." *Southern Economic Journal,* 41, 3 (January 1975): 347-381.

Gibbons, John H. "Energy: The Environmental Side." In The Conservation Foundation, *Energy Conservation Training Institute.* Washington, D.C.: The Conservation Foundation, n.d., pp. 1-33 - 1-76.

Gil, Efraim. *Energy Efficient Planning: An Annotated Bibliography.* PAS Report No. 315. Chicago: American Society of Planning Officials, March 1976.

Gilmore, Jack, and Duff, Mary. *Boom Town Growth Management.* Boulder, Col.: Westview Press, Inc., November 1975.

Gilmore, John S. "Boom Towns May Hinder Energy Resource Development." *Science,* 191, 4227 (February 13, 1976): 535–540.

Grier, Eunice S. "Changing Patterns of Energy Consumption and Costs in U.S. Households." Washington, D.C.: The Washington Center for Metropolitan Studies, September 18, 1976.

Grier, Eunice S. "National Survey of Household Activities." Washington, D.C.: The Washington Center for Metropolitan Studies, December 1975.

Hall, Shawn A., and Harrje, David T. *Instrumentation for the Omnibus Experiment in Home Energy Conservation.* Report 21. Princeton, N.J.: Center for Environmental Studies, Princeton University, May 1975.

Hannon, Bruce M. "An Energy Standard of Value." *The Annals of the American Academy,* 410 (November 1973): 139–153.

Hannon, Bruce M. "Energy and Labor Demand in the Conserver Society." Urbana: Energy Research Group, Center for Advanced Computation, University of Illinois at Urbana-Champaign, July 1976.

Hannon, Bruce M. "Energy Conservation and the Consumer." *Science,* 189 (July 11, 1975): 95–102.

Hannon, Bruce M. "Energy, Growth and Altruism." Urbana: Center for Advanced Computation, University of Illinois at Urbana-Champaign, October 21, 1975.

Hannon, Bruce M. "Energy, Land and Equity." In *Forty-First North American Wildlife Conference.* Washington, D.C.: Wildlife Management Institute, March 1976, pp. 256–276.

Hannon, Bruce M. "Independence Through Energy Conservation." Urbana: Energy Research Group, Center for Advanced Computation, University of Illinois at Urbana-Champaign, January 1975.

Hannon, Bruce M. "Energy, Labor, and the Conserver Society." *Technology Review* 79, 5 (March-April, 1977): 47–53.

Hanson, M.E., and Mitchell, J. W. *A Model of Transportation Energy Use in Wisconsin: Demographic Considerations and Alternative Scenarios.* IES Report 57. Madison: Institute for Environmental Studies, University of Wisconsin-Madison, December 1975.

Harrington, Winston. *Energy Conservation: A New Function for Local Governments?* Chapel Hill: Center for Urban and Regional Studies, University of North Carolina at Chapel Hill, December 1976.

Harrington, Winston. "Where Do Local Governments Fit Into an Energy Conservation Strategy?" *Carolina Planning,* 3 (Winter 1977): 43–52.

Harrje, David T. *Retrofitting: Plan, Action, and Early Results Using the Townhouses at Twin Rivers.* Report 29. Princeton, N.J.: Center for Environmental Studies, June 1976.

Harrje, David T., Junt, Charles M.; Treado, Steven J.; and Malik, Nicholas J. *Automated Instrumentation for Air Infiltration Measurements in Buildings.* Report 13. Princeton, N.J.: Center for Environmental Studies, Princeton University, April 1975.

Harwood, Corbin Crews. "Land Use and Energy Conservation." In the Conservation Foundation. *Energy Conservation Training Institute.* Washington, D.C.: The Conservation Foundation, n.d., pp. IV-165 – IV-184.

Harwood, Corbin Crews. *Using Land to Save Energy.* Cambridge, Mass.: Ballinger Publishing Company, 1977.

Healy, Robert G., and Hertzfield, Henry R. "Energy Conservation Strategies." In The Conservation Foundation, *Energy Conservation Training Institute.* Washington, D.C.: The Conservation Foundation, n.d., pp. IV-1 - IV-50.

Hirst, Eric. "Transportation Energy Conservation Policies." *Science,* 192 (April 2, 1976): pp. 15-20.

Hirst, Eric, and Carney, Janet. "Effects of Federal Residential Energy Conservation Programs," *Science,* 199 (February 24, 1978): pp. 845-851.

Hittman Associates, Inc. "Comprehensive Community Planning for Energy Management and Conservation." Interim Report. Washington, D.C.: Energy Research and Development Administration, April 1977.

Hittman Associates, Inc. *Residential Energy Consumption Multifamily Housing: Final Report.* U.S. Department of Housing and Urban Development Report HUD-HAI-4. Washington, D.C.: U.S. Government Printing Office, June 1974.

Hittman Associates, Inc. *Residential Energy Consumption in Single-Family Housing: Final Report.* U.S. Department of Housing and Urban Development Report HUD-HAI-2. Washington, D.C.: U.S. Government Printing Office, September 1975.

Keyes, Dale L. "Energy and Land Use: An Instrument of U.S. Conservation Policy?" *Energy Policy* (September 1976): 225-236.

Keyes, Dale L. "Land Use Control as a U.S. Policy Instrument for Conserving Energy." Washington, D.C.: The Urban Institute, 1975.

Keyes, Dale L. "Urban Form and Energy Use." Land Use Center Working Paper 9-5049-02. Washington, D.C.: The Urban Institute, September 1975.

Keyes, Dale L., and Peterson, George R. "Metropolitan Development and Energy Consumption." Land Use Center Working Paper 5049-15. Washington, D.C.: The Urban Institute, March, 1977.

Knecht, R. L., and Bullard, C. W. "End Uses of Energy in the U.S. Economy, 1967." Cal Document 145. Urbana: Center for Advanced Computation, University of Illinois at Urbana-Champaign, 1975.

Knowles, Ralph L. *Energy and Form: An Ecological Approach to Urban Growth.* Cambridge, Mass.: The MIT Press, 1974.

Large, David B. "Hidden Waste." In The Conservation Foundation, *Energy Conservation Training Institute.* Washington, D.C.: The Conservation Foundation, n.d., pp. 11-1 - 11-109.

Light, Alfred R. "Federalism and the Energy Crisis: A View from the States." *Publius,* The Journal of Federalism, 6 (Winter 1975): 81-96.

Light, Alfred R. "The Issue-Attention Cycle: North Carolina Looks at the Energy Crisis." *Public Affairs Forum,* IV (April 1975): pp. 1-6.

Liner, Gaines H. "A Methodology for Estimating Fiscal Impacts of Energy Price Changes on State and Local Government Outlays for Purchases of Goods and Services." Working Paper 77-WPIA-09. Washington, D.C.: Federal Energy Administration, March 1977.

Loebl, A. S.; Bjornstad, D. J.; Burch, D. F.; Howard, E. B.; Hull, J. F.; Madewell, D. G.; Malthouse, N. S.; and Ogle, M. C. *Transportation Energy Conservation Data Book.* ORNL-5198. Prepared by Oak Ridge National Laboratory for the Data Analysis Branch, Nonhighway Transport Systems and Special Projects,

Transportation Energy Conservation Division, Office of Conservation, Energy Research and Development Administration. Springfield, Va.: National Technical Information Service, October 1976.

Macrakis, Michael S., ed. *Energy: Demand, Conservation, and Institutional Problems*, Proceedings of a Conference Held at MIT. Cambridg, Mass.: The MIT Press, 1974.

Mayer, Lawrence S., and Robinson, Jeffrey A. *A Statistical Analysis of the Monthly Consumption of Gas and Electricity in the Home*. Report 18. Princeton, N.J.: Center for Environmental Studies, Princeton University, April 1975.

McFarland, William F. *Energy Development and Land Use in Texas*. Austin: Texas Governor's Energy Advisory Council, January 1975.

Myers, Phyllis. "Land-Use Policies and Energy Conservation." In The Conservation Foundation, *Energy Conservating Training Institute*. Washington, D.C.: The Conservation Foundation, n.d., pp. 111-73 - 111-106.

Moss, Laurence I. "Energy Conservation in the U.S.: Why? How Much? By What Means?" In The Conservation Foundation, *Energy Conservation Training Institute*. Washington, D.C.: The Conservation Foundation, n.d., pp. 1-1 - 1-32.

Mittal, Ram K., ed. *Effects of Energy Constraints on Transportation Systems*. Proceedings of the Third National Conference, held at Union College, Schenectady, N.Y., August 2-6, 1976. Prepared for Transportation Energy Conservation Division, Office of Conservation, Energy Research and Development Administration. Washington, D.C.: U.S. Government Printing Office, May 1977.

National League of Cities. "National Municipal Policy on Energy." Adopted by the Board of Directors, July 24, 1977. Washington, D.C.: National League of Cities, July 24, 1977.

National Research Council, Assembly of Behavioral and Social Sciences, Committee on Measurement of Energy Consumption. *Energy Consumption Measurement: Data Needs for Public Policy*. Washington, D.C.: National Academy of Sciences, 1977.

National Research Council, Transportation Research Board. *Transportation Energy Conservation and Demand*. Transportation Research Record 561. Washington, D.C.: Transportation Research Board, National Academy of Sciences, 1976.

National Research Council, Transportation Research Board, Commission on Sociotechnical Systems. *Scope and Plan for a Study of Transportation Energy Research Needs and Priorities: A Report to the Energy Research and Development Administration*. Washington, D.C.: Transportation Research Board, National Academy of Sciences, 1976.

Newman, Dorothy K., and Day, Dawn. *The American Energy Consumer*. Cambridge, Mass.: Ballinger Publishing Company, 1975.

Noll, Roger C. "Information, Decision-Making Procedures, and Energy Policy." *American Behavioral Scientist*, 19 (January-February 1976): 267-285.

Ogden, K. W. "The Effects of Different Forms of Urban Growth on Travel Patterns." In *Proceedings of the 5th Conference of the Australian Road Research Board*, vol. 5, 1970.

Orange County Energy Conservation Task Force. *Energy in Orange County*. Hillsborough, N.C.: The Task Force, 1976.

Pauker, Guy J. "Can Land Use Management Reduce Energy Consumption for

Transportation?" Rand Paper P-5241. Santa Monica, Calif.: Rand Corporation, May 1974.

Peskin, Robert L., and Schofer, Joseph L. *The Impacts of Urban Transportation and Land Use Policies on Transportation Energy Consumption.* Springfield, Va.: National Technical Information Service, April 1977.

Peskin, Robert L., and Schofer, Joseph L. "Urban Transportation and Land Use Policies and Energy Conservation." Paper prepared for the Office of University Research, Department of Transportation, under contract number USDOT-OS-50118. Evanston, Ill.: Department of Civil Engineering, Northwestern University, n.d.

Priest, W. Curtiss, Happy, Kenneth W., and Walters, Jeffrey L. *An Overview and Critical Evaluation of the Relationship Between Land Use and Energy Conservation.* Vols. I and II. Cambridge, Mass.: Technology + Economics, Inc., 1976.

Purdy, B. J., et al. "A Post Licensing Case Study of Community Effects at Two Operating Nuclear Power Plants. Draft. Oak Ridge, Tenn.: Oak Ridge National Laboratory, 1976.

Rastatter, Clem L., and Brinch, Jeannette. "Creative State Approaches to Energy Conservation." In The Conservation Foundation, *Energy Conservation Training Institute.* Washington, D.C.: The Conservation Foundation, n.d., pp. IV-235 - IV-279.

Real Estate Research Corporation. *The Costs of Sprawl.* Report prepared for the Council on Environmental Quality; the Office of Policy Development and Research, Department of Housing and Urban Development; and the Office of Planning and Management, Environmental Protection Agency. Washington, D.C.: U.S. Government Printing Office, April 1974.

Rechel, Ralph. "Federal and State Influences on Transportation Facilities, Services, and Fuel Consumption." In The Conservation Foundation, *Energy Conservation Training Institute.* Washington, D.C.: The Conservation Foundation, n.d., pp. 111-1 - 111-72.

Revelle, Randy. "Energy Conservation in Seattle." Paper prepared for Testimony before the United States House of Respresentatives, Committee on Banking, Finance and Urban Affairs, Subcommittee on the City, Washington, D.C., September 15, 1977.

Rice, Richard A. "System Energy and Future Transportation." *Technology Review,* 74 (January 1972): 31-37.

Rice, Richard A. "Toward More Transportation with Less Energy." *Technology Review,* 76 (February 1974): 44-53.

Richmond Regional Planning District Commission. *The Energy-Fuel Shortage and Land Development Trends in the Richmond Metropolitan Area: A Survey — January, 1974.* Washington, D.C.: Clearinghouse for Federal Scientific and Technical Information, RRPDC-CP-4, 1974.

Roberts, James S. "Energy and Land Use: Analysis of Alternative Development Patterns." *Environmental Comment* (September 1975): pp. 1-11.

Roberts, James S. *Energy, Land Use, and Growth Policy: Implications for Metropolitan Washington.* Washington, D.C.: Metropolitan Washington Council of Governments, June 1975.

Romanos, Michael. "Energy Conservation through Land Use Planning: A Framework for Research." Paper presented at the 23rd North American Meetings of the Regional Science Association, Toronto, November 1976.

Schneider, Jerry B., and Beck, J. R. "Reducing the Travel Requirements of the American City: An Investigation of Alternative Spatial Structure." Seattle: Department of Urban Planning and Civil Engineering, University of Washington, August 1973.

Shelley, Edwin F. "We Can Solve the Energy Crisis: The Transfer of Energy Conservation Technology." Paper presented at the 1976 Summer Workshop on Energy Extension Services at the University of California, Lawrence Berkeley Laboratory, July 19, 1976.

Shonka, D. B.; Loebl, A. S.; Ogle, M. C.; Johnson, M. L.; and Howard, E. B. *Transportation Energy Conservation Data Book: Edition I.5.* CONS/7405-1. Prepared by Oak Ridge National Laboratory for the Data Analysis Branch, Non-highway Transport Systems and Special Projects, Transportation Energy Conservation Division, Office of Conservation, Energy Research and Development Administration. Washington, D.C.: U.S. Government Printing Office, 1977.

Skidmore, Owings & Merrill. "Energy and Land Use." Portland, Ore.: Portland City Planning Commission, Fall 1976.

Smith, Richard B. "Energy Adjustments of Households." *Housing Educators Journal*, 3 (1976): 29-34.

Socolow, R. H., ed. *Saving Energy in the Home, Princeton's Experiment at Twin Rivers.* Cambridge, Mass.: Ballinger Publishing Company, 1978.

Socolow, Robert H. *Energy Utilization in Townhouses in a Planned Community in the United States.* Report 26. Princeton, N.J.: Princeton University Center for Environmental Studies, March 1976.

Socolow, Robert H., Harrje, David T., Mayer, Lawrence, and Seligman, Clive. *Energy Conservation in Housing: Work in Progress and Plans* for 1975-76. Report 19. Princeton, N.J.: Center for Environmental Studies, Princeton University, April 1975.

Southeastern Federal Regional Council. *Energy Handbook for State and Local Officials.* Atlanta: Southeastern Federal Regional Council, July 1976.

Stanford Research Institute. *Patterns of Energy Consumption in the United States.* Washington, D.C.: Government Printing Office, January 1972.

Swidler, Joseph C. "The Challenge to State Regulation Agencies: The Experience of New York State." *The Annals of the American Academy*, 410 (November 1973): 106-119.

Tether, Ivan. "Government Procurement and Energy Conservation." In The Conservation Foundation, *Energy Conservation Training Institute.* Washington, D.C.: The Conservation Foundation, n.d., pp. IV-111 - IV-130.

Thompson, Grant P. *Building to Save Energy—Legal & Regulatory Approaches.* Cambridge, Mass.: Ballinger Publishing Company, 1978.

Thompson, Grant P. "The Role of the States in Energy Conservation in Buildings." In The Conservation Foundation, *Energy Conservation Training Institute.* Washington, D.C.: The Conservation Foundation, n.d., pp. IV-215 - IV-234.

U.S. Bureau of the Census, *Detailed Housing Characteristics 1970* Washington, D.C.: U.S. Government Printing Office, 1972.

United States Congress, House of Representatives, Committee on Banking, Finance and Urban Affairs. *Energy and the City.* Hearings before the Subcommittee on the City, Ninety-fifth Congress, First Session, September 14-16, 1977. Washington, D.C.: U.S. Government Printing Office, 1977.

United States Congress, Office of Technology Assessment. *An Analysis of* the *ERDA Plan and Program.* Washington, D.C.: U.S. Government Printing Office, October, 1975.

United States Congress, Office of Technology Assessment. *Energy, The Economy, and Mass Transit.* Washington, D.C.: U.S. Government Printing Office, December 1975.

United States Congress, Senate Committee on Interior and Insular Affairs. *Land Use and Energy: A Study of Interrelationships.* Report. Prepared by Environmental Natural Resources Policy Division, Congressional Research Service, Library of Congress. Washington, D.C.: U.S. Government Printing Office, January 1976.

U.S. Department of Commerce, Social and Economic Statistic Administration, Bureau of Economic Analysis. *Input-Output Structure of the U.S. Economy;* 1967. Vols. 1, 2, 3. Washington, D.C.: U.S. Government Printing Office, 1974.

U.S. Department of Commerce, National Bureau of Standards. *Technical Options for Energy Conservation in Buildings.* Washington, D.C.: U.S. Government Printing Office, July 1973.

U.S. Department of Commerce, Office of Business Economics. *Input-Output Structure of the U.S. Economy: 1963.* Vols. 1, 2, 3. Washington, D.C.: U.S. Government Printing Office, 1969.

U.S. Department of Housing and Urban Development, Office of Community Planning and Development. *Rapid Growth From Energy Projects: Ideas for State and Local Action.* Washington, D.C.: U.S. Government Printing Office, 1976.

U.S. Department of Transportation, Urban Planning Division, Office of Highway Administration, Federal Highway Administration. *Energy Conservation in Ground Transportation: A Comparison of Alternative Strategies.* Washington, D.C.: U.S. Department of Transportation, August 1977.

Watt, Kenneth E. F., and Ayers, Claudia. "Urban Land Use Patterns and Transportation Energy Cost." Davis: University of California at Davis, 1974.

Williams, Robert H., ed. *The Energy Conservation Papers.* Cambridge, Mass.: Ballinger Publishing Company, 1975.

Willman, Phillip. "The Consumer's Choice." In The Conservation Foundation, *Energy Conservation Training Institute.* Washington, D.C.: The Conservation Foundation, n.d., pp. V-1 – V-36.

Zaelke, Durwood J. "Energy Conservation and Urban Transportation." In The Conservation Foundation, *Energy Conservation Training Institute.* Washington, D.C.: The Conservation Foundation, n.d., pp. IV-131 – IV-164.

Zehner, Robert B. *Access, Travel, and Transportation in New Communities.* Cambridge, Mass.: Ballinger Publishing Company, 1977.

Index

About the Authors

T. Owen Carroll is Associate Professor in the W. Averell Harriman College, the State University of New York at Stony Brook. He has played a major role, along with researchers in the Department of Applied Science at Brookhaven National Laboratory, in the development of a complex model to simulate land use-transportation-energy interactions. Carroll is co-author of *The Planner's Energy Workbook: A User's Manual for Exploring Land Use and Energy Utilization Relationships, Land Use-Energy Simulation Model: A Computer-Based Model for Exploring Land Use and Energy Relationships*, and a number of other articles and reports on energy and land use.

John H. Cumberland is Professor of Economics and Director of the Bureau of Business and Economic Research, The University of Maryland. He has recently developed a long-range schedule for construction of new electric power plants in Maryland, which attempts to meet predicted demands while minimizing environmental impacts. Cumberland, who contributed to the economic and environmental analysis of the Calvert Cliffs, Maryland, Nuclear Plant, has also developed a statewide environmental accounting system and a state economic-environmental planning model. He is the author of "Forecasting Electric Energy Requirements and Environmental Impacts for Maryland, 1970–1990," and "Interdependence Between Energy, Environment, Regional Development, and Economic Growth."

Jerry L. Edwards is Assistant Professor in the Department of Civil and Mineral Engineering, The University of Minnesota. Involved for several years in the use of computer simulation to compare the transportation energy efficiency of alternative urban forms, he was funded by the National Science Foundation to examine the transportation energy implications of establishing subregional commercial and governmental centers in the Minneapolis-St. Paul area. Edwards is the author of "Relationships Between Transportation Energy Consumption and Urban Spatial Structure," "Use of a Lowry-Type Spatial Allocation Model in an Urban Transporation Study," and "Relationships between Transportation Energy Consumption and Urban Structure: Results of Simulation Studies."

Kim Gillan is an energy analyst with the Washington Office of the State of Texas. Her contribution to this volume was prepared while she was Program Manager with the Division of Buildings and Community Systems of the Energy Research and Development Administration's Office of Conservation. Gillan played a major role in the development of the federal energy research program in buildings and community design, including the solicitation of new research proposals and the evaluation of ongoing projects.

John S. Gilmore is Senior Research Economist with the Denver Research Institute, University of Denver. He has done extensive work on the impacts of power plants and other energy development projects in the Rocky Mountain and Great Plains states of Wyoming, Colorado, North Dakota, Utah, and Montana. Gilmore is the author of *Boom Town Growth Management* and "Boom Towns May Hinder Energy Resource Development."

Eunice S. Grier, a private consultant in Bethesda, Maryland, was formerly affiliated with the Washington Center for Metropolitan Studies. She has also served as Research Director of the U.S. Commission on Civil Rights. In 1975, she directed a national household survey, sponsored by the Federal Energy Administration, to provide data that could be compared with the 1973 national survey conducted for the Ford Foundation's Energy Policy Project. Grier has recently completed a study of the impacts of energy shortages and price increases on low-income households for the Community Service Administration, *Colder . . . Darker! The Energy Crisis and Low Income Americans.* She is the author of "Changing Patterns of Energy Consumption and Costs in U.S. Households" and co-author of "Natural Gas Usage and Consumption by American Households" and *Equality and Beyond: Housing Segregation and the Goals of the Great Society.*

Bruce Hannon is Associate Professor in the Center for Advanced Computation of the University of Illinois at Urbana-Champaign and Director of the center's Energy Research Group. His current research includes studies of energy use for building construction and the energy conservation and employment impacts of changes in technology and consumption. A recipient of the Mitchell Prize for his paper "Energy, Growth and Altruism," Hannon is also the author of "Energy, Labor and the Conserver Society," "Independence through Energy Conservation," "Energy, Land and Equity," "Energy Conservation and the Consumer," and "An Energy Standard of Value."

David T. Harrje, Senior Research Engineer with the Center for Environmental Studies of Princeton University, is codirector of a detailed study of household energy consumption funded originally by the National Science Foundation and more recently by the Department of Energy. The study team has used a combination of sophisticated instruments and behavioral research methods to evaluate the energy consumption patterns and energy conservation potentials of a group of identical occupied townhouses in Twin Rivers, New Jersey. Harrje is the author of *Retrofitting: Plan, Action, and Early Results Using the Townhouses at Twin Rivers* and co-author of *Instrumentation for the Omnibus Experiment in Home Energy Conservation* and *Automated Instrumentation for Air Infiltration Measurements in Buildings.*

Stephen D. Julias, III is Vice President in the Development Research Division of Booz, Allen & Hamilton, Inc. He recently directed an extensive empirical study of the energy conservation potential in existing metropolitan communities for the Federal Energy Administration. The report, *Interaction of Land Use Patterns and Residential Energy Conservation,* describes his team's findings.

Dale L. Keyes is Environmental Policy Analyst with the firm, Energy and Environmental Analysis, Arlington, Virginia. His contribution to this volume was prepared while he was Research Associate with the Urban Institute in Washington, D.C. In addition to his research on energy and land use, Keyes has conducted research in air pollution control and on the environmental ramifications of metropolitan development. Keyes is the author of "Energy and Land Use: An Instrument of U.S. Conservation Policy?" and co-author of *Metropolitan Development and Energy Consumption.*

James S. Roberts is Principal Counselor for the Real Estate Research Corporation in Chicago. He is presently involved in developing methods for community energy planning, as part of a study funded by the Energy Research and Development Administration. Roberts is the author of "Energy and Land Use: Analysis of Alternative Development Patterns," *Energy, Land Use, and Growth Policy: Implications for Metropolitan Washington,* and co-author of *The Costs of Sprawl.*

Michael Sizemore heads the architectural and energy planning firm of Sizemore & Associates in Atlanta, Georgia. He has extensive experience in the design and use of solar heating systems. His recent projects include a study for the Energy Research and Development Administration of the energy conservation potential of neighborhood in-filling; the design of an energy-conserving neighborhood of rehabilitated, elderly, and new housing, and recreation areas (sponsored by the National Endowment for the Arts); retrofitting of large office buildings to reduce energy consumption; and the development of a methodology for the creation of energy management plans suitable for use by small communities that lack extensive equipment, resources, data, or expertise.

Stephen S. Skjei is Resource Environmental Economist with the Site Designation Standards Branch, Office of Standards Development, United States Nuclear Regulatory Commission. He is involved in evaluating the economic and environmental impacts of nuclear power plants, and in developing standardized methodological approaches to power plant siting questions.

Grant P. Thompson is a Senior Associate at The Conservation Foundation, Washington, D.C. He formerly headed the energy research program at the Environmental Law Institute and served as an Energy Specialist with the Pacific Northwest Regional Commission. He has directed numerous energy conservation research projects, including several dealing with the land use implications of solar energy. He is currently serving on a Ford Foundation-sponsored study group, under Resources for the Future, entitled, "Energy: The Next Twenty Years." Thompson is the author of "Design Features, Shading, and Orientation for Energy Conservation," "The Role of the States in Energy Conservation in Buildings," *Building to Save Energy: Legal & Regulatory Approaches,* and co-author of *Energy and the Social Sciences.* He has been a contributor to the *1974 Municipal Yearbook, Public Management,* and *Federal Environmental Law.* He also served in 1976 as a Contributing Editor of the *ASHRAE Journal*, writing on energy conservation and the law.

About the Editors

Raymond J. Burby, III is Assistant Director for Research at the Center for Urban and Regional Studies of The University of North Carolina at Chapel Hill and has directed the center's research program on energy and patterns of human settlement. He received the M.R.P. and Ph.D. in planning from The University of North Carolina at Chapel Hill. He is the author of numerous articles and books on community development, including *Recreation and Leisure in New Communities* and *Planning and Politics: Toward a Model of Planning-Related Policy Outputs in American Local Government*; co-author of *New Communities U.S.A.*, *Health Care in New Communities*, and *Schools in New Communities*, and co-editor of *New Community Development: Planning Process, Implementation, and Emerging Social Concerns*.

A. Fleming Bell is an urban planner with the Rockingham-Richmond County, North Carolina, planning department. His work on this volume was undertaken while he served as Energy Research Assistant for the Center for Urban and Regional Studies, The University of North Carolina at Chapel Hill. Bell, who received the M.R.P. in Planning from The University of North Carolina, has also worked for the Old Colony Planning Council, Brockton, Massachusetts, and for the American Council on Education. His special interests include flood plain management, the benefits and costs of urban erosion and sedimentation control programs, and the problems caused by rapid growth in small towns and rural areas.